The Disposal and Utilisation of Abattoir Waste in the European Communities

Werner Weiers and Roland Fischer

A report prepared for the Environment and Consumer Protection
Service of the Commission of the European Communities
by Werner Weiers and Roland Fischer

Springer-Science+Business Media, B.V.

for the Commission of the European Communities, Directorate-General Scientific and Technical Information and Information Management, Luxembourg

© Springer Science+Business Media Dordrecht 1978
Originally published by Graham & Trotman Limited in 1978

ISBN 978-94-009-9654-0   ISBN 978-94-009-9652-6 (eBook)
DOI 10.1007/978-94-009-9652-6

Contents

List of tables

List of figures

1 Introduction

The problem of pollution of the environment arose for the first time
with the beginning of industrialization.

At that time there was no need to think about disposal of waste. There
was no reason to do so because the space available to man and other
living creatures appeared quite adequate. The life expectancy of man
has risen as a result of the enormous advances in technology and medicine.
In 1650 it was 30 years throughout the world while in 1970 it had risen
to 53 years, and this has consequently been accompanied by a population
explosion.

While the world population in 1650 was some 500,000,000, in 1970 it was
about 3,600,000,000.

Assuming that mortality does not rise sharply, the population potential
for the year 2000 is expected to be some 7,000,000,000[1].

Industrial production has grown to an even greater degree than the
world population.

The growth of noxious substances has developed with the growth of the
population on the one hand and with industrial growth on the other.

The growth of slaughterhouse waste is directly related to population
growth, because as the population increases so does the demand for meat
and meat products. Consumption habits have also changed with increasing
prosperity so that there has been an increase in the proportion of
slaughterhouse by-products which are difficult to market.

Before the Second World War slaughtering was carried out mainly in
small to medium-size butchers' undertakings, but soon after the structure
of butchers' undertakings and slaughterhouses began to change.

Slaughterhouses and meat distributors began to grow in size and number.
This development has had a positive effect on the problem of slaughterhouse waste, because it has been possible to deal with and dispose of
such waste more satisfactorily. Since, however, most slaughterhouses,
and particularly the older ones, regard the problem of waste disposal
as of secondary importance, there are still considerable problems in
respect of environmental improvement in slaughterhouses.

The purpose of this study is as follows:

1. To analyze the slaughterhouse waste occurring in the Member States
    of the Community.

2. To analyze the present disposal of slaughterhouse waste in the
    Member States.

3. To project the incidence of slaughterhouse waste in the EEC over
    the next 10 years and the next 20 years.

4. To draft proposals for possible Community measures for future
    disposal of slaughterhouse waste in the Community and for the use
    and utilization of slaughterhouse waste in industry and agriculture
    from economic and environmental aspects.

2 Definition of 'slaughter house waste'

The term 'slaughterhouse waste' generally refers to all waste occurring in a slaughterhouse as a result of slaughtering and the waste resulting from the operation of a slaughterhouse.

It is therefore divided up into the following:

1. Slaughtering waste (carcass waste pure and simple)
2. Other waste

Slaughtering waste is the waste which is not used for human consumption either because it is unfit or because it is refused by the consumer for reasons associated with consumer habits.

Slaughtering waste also includes condemned meat and carcasses. Condemned meat is confiscated carcasses and parts thereof.

The *other waste* includes waste which occurs at slaughterhouses as a result of the animals having to be stabled some days before slaughtering. This waste comprises faeces, urine and litter (dung).

Waste water is another problem. Considerable quantities of waste (blood, contents of stomach and intestines, bristles, etc) are disposed of via the waste water system at many abattoirs. The author has confirmed such findings on visiting a number of abattoirs.

Slaugherhouse waste comprises the following:

    Carcass waste
    + condemned meat
    + animal excrement and litter
    + waste water
    —————————————————
    = total slaughterhouse waste

3 Determination of quantities of slaughterhouse waste in the EEC Member States

The individual statistics are confined to indicating meat production and edible offal. The inedible slaughterhouse waste is not listed separately in the statistics of the individual countries.

Meat production figures are based on the numbers of livestock in the individual EEC Member States as listed in table I for the years 1970 to 1971.

Table 2 shows the number of animals slaughtered in the individual EEC Member States and relates to both home-raised and imported animals for the years 1970 to 1973.

Table 3 shows the net meat production broken down over the individual EEC Member States, the meat being defined as the carcass weight of the animal after drawing, including the bones. This net production relates to meat from animals slaughtered in the country concerned, ie home-raised animals and imported animals killed.

Table 1: EEC livestock (production basis)

By countries: Units: '000s

Country	Year	Cattle	Pigs	Chickens	Horses	Sheep	Goats	Donkeys
FRG	1970	14,026	20,969	98,601	253	843	50	—
	1971	13,638	19,985	99,530	265	850	43	—
	1972	13,892	20,028	99,712	283	908	40	—
	1973	14,360	20,394	—	—	—	—	—
France	1970	21,737	11,215	105,000*	629	10,239	924	33
	1971	21,764	11,279	105,000*	524	10,115	909	29
	1972	22,509	11,375	105,000*	480	10,191	899	30
	1973	23,949	11,454	—	—	—	—	—
Italy	1970	8,776	8,980	—	271	7,948	1,019	263
	1971	8,669	8,196	—	255	7,846	976	235
	1972	8,818	7,996	—	258	7,805	962	214
	1973	8,407	8,201	—	—	—	—	—
Netherlands	1970	3,865	6,340	55,375	59	375	—	—
	1971	3,748	6,316	60,125	50	375	—	—
	1972	4,111	6,480	58,430	48	388	—	—
	1973	4,978	6,425	—	—	—	—	—

Table 1 continued

Country	Year	Cattle	Pigs	Chickens	Horses	Sheep	Goats	Donkeys
Belgium	1970	2,715	3,835	--	67	66	3	--
	1971	2,643	3,925	--	60	66	3	--
	1972	2,750	4,298	--	58	--	--	--
	1973	2,975	4,620	--	--	--	--	--
Luxembourg	1970	186*	131	370*	2	5	--	--
	1971	186*	109	355*	2	5	--	--
	1972	192	103	297	2	5	--	--
	1973	192*	103*	--	--	--	--	--
UK	1970	12,442	8,546	135,825	136	--	--	--
	1971	12,928	8,882	137,280	134	--	--	--
	1972							
	1973	14,782	9,215	--	--	--	--	--
Ireland	1970	5,405	1,155	9,341	124	--	--	--
	1971	5,516	1,144	9,500	124	--	--	--
	1972							
	1973	6,406	1,034	--	--	--	--	--
Denmark	1970	2,842	8,361	17,847	45	--	--	--
	1971	2,723	8,626	16,220	47	--	--	--
	1972							
	1973	2,956	8,414	--	--	--	--	--

Table 2: Number of animals killed (home-raised and imported animals)
in the individual countries

By countries: Units: '000s

Country	Year	Cattle	Calves	Pigs	Poultry	Horses	Sheep and Goats
FRG	1970	4,628	1,068	29,220	21,521	15	482
	1971	4,674	1,053	31,196	22,676	15	508
	1972	4,007	890	30,850	23,440	15	559
	1973	4,062	741	30,506	--	15	533
France	1970	3,917	4,179	15,703	--	231	6,597
	1971	4,059	4,065	17,147	--	215	7,251
	1972	3,582	3,458	17,583	--	180	7,260
	1973	3,562	3,163	17,546	--	150	7,241
Italy	1970	3,903	1,287	6,383	--	316	6,588
	1971	3,892	1,238	6,933	--	299	6,678
	1972	3,836	1,044	7,142	--	342	6,672
	1973	3,890	1,060	7,341	--	319	6,280
Netherlands	1970	925	1,015	8,468	25,315	19	438
	1971	891	1,026	9,658	25,545	14	474
	1972	703	930	9,525	24,598	10	420
	1973	706	962	9,695	--	8	388

Table 2 continued

Country	Year	Cattle	Calves	Pigs	Poultry	Horses	Sheep and Goats
Belgium	1970	766	281	5,782	11,063	30	174
	1971	789	284	6,311	11,765	25	149
	1972	745	267	6,642	11,580	20	135
	1973	752	228	7,186	--	16	136
Luxembourg	1970	44	2	127	--	0	0
	1971	41	2	124	--	0	0
	1972	29	1	124	--	0	0
	1973	29	1	123	--	0	0
UK	1970	3,683	356	14,391	--	--	11,458
	1971	3,625	257	15,566	--	--	11,297
	1972	3,487	153	15,440	--	--	11,047
	1973	3,307	142	15,128	--	--	11,805
Ireland	1970	899	5	2,102	--	--	1,484
	1971	910	5	2,284	--	8	1,820
	1972	779	4	2,368	--	8	1,761
	1973	866	4	2,111	--	8	1,690
Denmark	1970	488	605	11,497	--	--	58
	1971	479	587	12,093	--	2	39
	1972	411	471	11,839	--	3	31
	1973	424	445	11,424	--	2	25

0 = less than 1,000

-- = no details

Source: (4) (5) (6)

Table 3: Net production* by country
Units: '000s

Country	Year	Cattle	Calves	Pigs	Poultry	Solipeds	Sheep and Goats
FRG	1970	1,274	83	2,583	269	5	11
	1971	1,295	81	2,737	275	4	13
	1972	1,126	75	2,731	264	4	13
	1973	1,169	66	2,681	293	4	14
France	1970	1,187	378	1,375	638	73	120
	1971	1,236	377	1,491	652	66	132
	1972	1,122	334	1,541	717	55	134
	1973	1,138	322	1,544	791	46	133
Italy	1970	954	123	593	631	51	55
	1971	964	120	644	659	52	54
	1972	948	101	666	725	56	52
	1973	961	112	689	806	53	50
Netherlands	1970	258	104	701	292	5	11
	1971	242	106	795	319	3	12
	1972	199	100	789	328	3	11
	1973	200	108	812	339	2	10
Belgium	1970	230	27	465	110	10	4
	1971	238	28	498	112	8	4
	1972	231	27	532	114	7	3
	1973	236	23	580	116	5	3

Table 3 continued

	Year	Cattle	Calves	Pigs	Poultry	Solipeds	Sheep and Goats
Luxembourg	1970	10	0	10	-	--	0
	1971	9	0	10	-	--	0
	1972	8	0	9	-	--	0
	1973	8	0	9	-	--	0
UK	1970	938	11	920	--	-	227
	1971	929	8	983	589	-	225
	1972	898	6	980	679	-	220
	1973	850	6	978	663	--	232
Ireland	1970	216	0	144	--	--	40
	1971	236	0	154	35	2	47
	1972	204	0	160	40	2	45
	1973	207	0	145	43	2	43
Denmark	1970	113	77	738	--	--	2
	1971	115	79	764	80	-	2
	1972	101	70	765	85	-	1
	1973	106	77	774	90	3	-

*Net production = meat of animals slaughtered in the country concerned (meat from home-raised and imported animals

Meat = carcass weight of the animal after drawing, including bones

Source: (4)

O = insignificant

Schön, Holz and Belendorff developed coefficients specific to each type of animal for calculating waste quantities for the individual disposal system[2].

The disposal systems listed are waste water, special utilization and carcass disposal institutions (TBA). These animal-specific coefficients relate to the quantity of meat produced.

Table 4 gives these animal-specific coefficients.

Table 4: Animal-specific coefficients for calculating waste quantities for the individual disposal systems

Type of animal	Waste water	Special utilization	TBA
Calves	0.093	0.036	0.086
Cattle	0.256	0.386	0.122
Pigs	0.077	0.172	0.107
Sheep	0.149	0.379	0.119
Horses	0.120	0.292	0.243
Poultry	0.024	0.123	0.089

Source: (2)

Table 5 gives the slaughterhouse waste for 1973 broken down by the individual countries and by disposal systems determined on the basis of the coefficients. The calculation was based on the net production given in Table 3.

The following quantities of slaughterhouse waste were determined:

Federal Republic of Germany	1,946,000 t
France	1,803,000 t
Italy	1,275,000 t
Netherlands	433,000 t
Belgium	426,000 t

Luxembourg	9,240 t
UK	1,320,000 t
Ireland	249,000 t
Denmark	398,000 t

The quantities determined contain only the waste arising at slaughter-houses as a result of the slaughtering process.

The following must be added to this:

1. The quantities from animals which died on the way to the slaughterhouse or abattoir.

2. The quantities condemned by meat inspectors as unfit for consumption

14

Table 5: Slaughtering waste by countries, 1973 — Disposal systems and total quantity

Units: '000 t

Country	Disposal	Cattle	Calves	Pigs	Poultry	Solipeds	Sheep and Goats	Total
FRG	Waste water	299	6.1	206	12	0.4	2.0	526
	Special utilization	451	2.3	461	36	1.1	5.3	956
	TBA*	142	5.6	287	26	0.9	1.6	464
	Total quantity	892	14.0	954	74	2.6	9.0	1,946
France	Waste water	291	29.9	119	33	5.5	19.8	498
	Special utilization	439	11.5	266	97	13.4	50.4	877
	TBA	138	27.6	165	70	11.1	15.8	428
	Total quantity	868	69.0	550	200	30.0	86.0	1,803
Italy	Waste water	246	10.4	53	34	6.3	7.4	357
	Special utilization	370	4.0	119	99	15.4	18.9	626
	TBA	117	9.6	74	72	12.8	5.9	292
	Total quantity	733	24.0	246	205	34.5	32.2	1,275
Netherlands	Waste water	51	10.0	63	14	0.2	1.5	140
	Special utilization	77	3.8	14	42	0.6	3.8	141
	TBA	24	9.2	87	30	0.5	1.2	152
	Total quantity	152	23.0	164	86	1.3	6.5	433

Table 5 continued

Country	Disposal	Cattle	Calves	Pigs	Poultry	Solipeds	Sheep and Goats	Total
Belgium	Waste water	60	2.1	45	4.8	0.6	0.4	113
	Special utilization	91	0.8	100	14.2	1.4	1.1	208
	TBA	29	1.9	62	10.3	1.2	0.3	105
	Total quantity	180	4.8	207	29.3	3.2	1.8	426
Luxembourg	Waste water	2.0	-	0.7	0.04	-	-	2.74
	Special utilization	3.0	-	1.5	0.12	-	-	4.62
	TBA	0.9	-	0.9	0.08	-	-	1.88
	Total quantity	5.9	-	3.1	0.24	-	-	9.24
UK	Waste water	218	0.5	75	28	-	35	357
	Special utilization	328	1.0	168	82	-	88	667
	TBA	104	0.6	105	59	-	28	296
	Total quantity	650	2.1	348	169	-	151	1,320
Ireland	Waste water	53	-	11	1.8	0.24	6.4	72
	Special utilization	80	-	25	5.2	0.58	16.2	127
	TBA	25	-	16	3.8	0.48	5.1	50
	Total quantity	158	-	52	10.8	1.30	27.7	249
Denmark	Waste water	27	7.1	60	3.7	0.36	0.14	98
	Special utilization	41	2.7	133	11.0	0.87	0.37	189
	TBA	13	6.6	83	8.0	0.72	0.11	111
	Total quantity	81	16.4	276	22.7	1.95	0.62	398

Source: (2)

*TBA = carcass disposal institutions

1. _Animals dying on the way to the slaughterhouse_

Since no statistics are available, the following quantities are estimated for 1973:

Federal Republic of Germany	ca.	81,000 t
France	ca.	123,000 t
Italy	ca.	52,000 t
Netherlands	ca.	27,000 t
Belgium	ca.	16,000 t
Luxembourg	ca.	1,000 t
UK	ca.	83,000 t
Ireland	ca.	31,000 t
Denmark	ca.	20,000 t

2. _Meat condemned as unfit for human consumption_

It is difficult to determine these quantities, since no adequate statistics are available.

The following are the estimated quantities of condemned meat for 1973:

Federal Republic of Germany	ca.	8,000 t
France	ca.	6,500 t
Italy	ca.	6,800 t
Netherlands	ca.	1,500 t
Belgium	ca.	1,500 t
Luxembourg	ca.	60 t
UK	ca.	6,600 t
Ireland	ca.	1,700 t
Denmark	ca.	850 t

Since not all EEC Member States have strict meat inspection, these values can only be regarded as a guide.

The quantities of meat arising as a result of dead animals and condemned meat are handled by the carcass disposal institutions.

The following quantities apply to the total slaughtering waste for the individual countries.

Federal Republic of Germany	2,035,000 t
France	1,932,500 t
Italy	1,333,800 t
Netherlands	461,500 t
Belgium	443,500 t
Luxembourg	9,301 t
UK	1,409,600 t
Ireland	281,700 t
Denmark	418,850 t

These quantities have to be handled by various disposal systems.

Slaughterhouse waste includes not only waste from the slaughtering operation but also waste arising out of the stabling of the animals before slaughtering at the slaughterhouse. This waste is made up of large quantities of faeces, urine and litter.

It is impossible to determine these quantities because the animals are stabled at the slaughterhouse for varying periods before slaughtering and there are no data available. Also, most of this waste passes directly into the waste water.

Disposal of this type of waste gives rise to considerable problems for many slaughterhouses, particularly in urban areas. Unsuitable disposal of this type of waste may result in environmental nuisance and pollution.

4 Slaughterhouse waste broken down by type

A distinction should be made between the following types of slaughter-house waste:

Hide/skin	Pieces of hide, skin from head and legs, and sinews
Bristles/hair	
Claws	
Horns	
Lungs	
Spleen	
Udder	
Rumen	
Stomach	Stomach contents
Trachea	
Chitterlings	
Blood	
Intestines	Contents of intestines
Reticulum	
Bladder	
Feet	
Horn roots	
Bones from the carcass	
Pieces of fat	
Sex organs - testicles, penis, vagina, scrotum	
Eyes	
Ears	
Gall bladder	Bile
Glands	Pituitary gland, pineal gland, thyroid gland, pancreas, adrenal gland, ovaries and uterus

Carcasses and carcass parts unfit for consumption

Poultry waste	Gullet, trachea, intestine, lungs, sex organs, intestine contents, head, stomach contents, shanks and webbed feet, feathers

Stabling waste   Faeces, urine, litter

Waste water

The sex organs, eyes, ears and glands are compulsorily condemned.

Compulsorily condemned meat originates not only from sick animals but also from healthy animals and covers those parts of the carcass which are not released for human consumption.

Carcasses and carcass parts unfit for consumption are condemned parts which must not be used for human consumption for hygienic and health reasons and must be disposed of harmlessly.

5 Determination of slaughterhouse waste by type and quantity

The Individual types of slaughterhouse waste were listed In Chapter 4. In this chapter we shall try to determine the quantities of the Individual types.

Table 6 gives the proportions of waste from carcasses of cattle, calves, sheep, pigs and horses. These values have been taken from Vanselow's paper[3]. No literature exists for the other animals (donkeys and goats).

To enable the individual animals to be determined quantitatively, the average carcass weights of the individual animals in the individual EEC Member States must be determined. For this reason, table 7 lists the average carcass weights in kg for the years 1970 to 1973 broken down by country. The mean carcass weights have been calculated from these three years (see table 8).

The proportions of the individual types of waste have been calculated in tables 9 - 17, more particularly by reference to the average carcass weights given in table 8. The calculations have been carried out separately for each EEC country.

The waste is inedible condemned meat or waste which is not used for human consumption.

Table 6:   Proportion of waste from slaughtered carcasses

Units: Percentage of whole carcass weight

Type	Cattle	Calves	Sheep	Pigs	Horses
Hide/skin total	12.76	14.37	12.28	-	8.04
Rind	-	-	-	7.58	-
Bristles/hair	-	-	-	0.50	-
Claws	0.49	-	-	0.06	+
Horns	0.13	-	-	-	-
Lungs	1.52	1.63	3.75	0.57	2.34
Spleen	0.31	0.45	0.67	0.16	0.55
Udder	0.89	-	3.08	-	-
Stomach, empty	5.86	2.37	9.22	1.0	0.74
Stomach contents	20.00	7.45	+	0.53	+
Tranchea	0.31	0.69	+	0.35	+
Chitterling	0.32	2.81	+	0.42	+
Blood	5.12	5.84	6.78	4.03	9.08
Intestines, empty	3.20	5.30	4.57	4.30	8.23
Intestine contents	6.00	x	+	3.30	18.14
Reticulum and small intestine	1.79	1.20	4.57	1.40	-
Bladder	0.10	-	+	0.10	-
Bones from slaughtered carcasses	16.96	23.98	17.11	11.49	19.84
From horn roots/head	3.69	4.22	3.85	-	-
From feet	1.56	2.37	3.39	-	6.93
Fat tissue	5.44	-	-	15.20	4.92
Sex organs	1.09	-	+	0.35	+
Eyes	0.04	-	+	0.04	+
Gall bladder, empty	0.03	-	+	0.04	-
Bile	0.14	-	+	0.13	-
Ears	-	-	-	0.10	+
Other waste	-	-	25.23	-	3.0

x In the case of stomach contents   + In the case of other waste

Source: (3)

Table 7: Average carcass weights

Units : kg

Country	Year	Cattle	Calves	Pigs	Solipeds	Sheep and Goats
FRG	1970	275	77	89	305	23
	1971	277	78	89	297	26
	1972	281	84	89	308	23
	1973	288	89	88	303	24
France	1970	303	90	88	316	18
	1971	305	93	87	307	18
	1972	313	97	88	306	19
	1973	320	102	88	309	18
Italy	1970	244	96	93	158	8
	1971	248	97	91	162	8
	1972	247	97	93	164	8
	1973	247	106	94	161	8
Netherlands	1970	270	102	83	276	25
	1971	272	103	82	259	25
	1972	283	106	83	268	26
	1973	283	112	84	267	26

Table 7 continued

Country	Year	Cattle	Calves	Pigs	Solipeds	Sheep and Goats
Belgium	1970	302	94	80	329	23
	1971	302	97	79	320	27
	1972	310	91	80	331	22
	1973	304	97	81	326	24
Luxembourg	1970	265	86	75	..	..
	1971	268	100	73	..	..
	1972	276	112	73	..	..
	1973	269	99	73	..	..
	1970	254	31	63	..	19
	1971	256	31	63	..	20
	1972	263	33	63	..	20
	1973	258	42	65	..	20
UK	1970	240	-	68	-	27
	1971	259	91	67	250	26
	1972	262	91	68	250	26
	1973	239	91	68	250	25
Ireland	1970	231	127	64	-	34
	1971	240	135	63	500	35
	1972	246	149	65	333	35
	1973	250	173	68	-	35

Source: (4)

- : no evidence available

Table 8:   Average carcass weights by animal and country
            from 1970 to 1973

Units : kg

Country	Cattle	Calves	Pigs	Solipeds	Sheep and Goats
FRG	280	82	88	303	24
France	310	95	87	309	18
Italy	246	99	92	161	8
Netherlands	277	105	83	267	25
Belgium	304	94	80	326	24
Luxembourg	269	99	73	300	24
UK	257	34	64	270	20
Ireland	250	91	68	250	26
Denmark	241	146	65	290	35

The values for Luxembourg, Ireland and Denmark have been partly
estimated.

Source: Calculated from Table 7

Federal Republic of Germany

Table 9: Proportion of waste in absolute terms, based on average carcass weights shown in table 8

Units : kg

Type	Cattle	Calves	Pigs	Sheep	Horses
Hide	35.72	11.78	-	2.94	24.36
Rind	-	-	6.67	-	-
Bristles	-	-	0.44	-	-
Claws	1.37	-	0.05	-	+
Horns	0.36	-	-	-	-
Lungs	4.25	1.33	0.50	0.90	7.09
Spleen	0.86	0.36	0.14	0.16	1.66
Udder	2.49	-	-	0.74	-
Stomach, empty	16.40	1.94	0.88	2.21	2.24
Stomach contents	56.00	6.10	0.46	+	+
Trachea	0.86	0.56	0.30	+	+
Chitterlings	0.89	2.30	0.36	+	+
Blood	14.33	4.78	3.54	1.62	27.51
Intestines, empty	8.96	4.34	3.78	1.09	24.93
Intestine contents	16.80	x	2.90	+	54.96
Reticulum and small intestine	5.01	0.98	1.23	1.09	-
Bladder	0.28	-	.0.08	+	-
Bones from slaughtered carcasses	47.48	19.66	10.11	4.10	60.86
From horn roots/head	10.33	3.46	-	-	-
From feet	4.36	-	-	3.39	20.99
Fat tissue	15.23	1.94	13.37	-	14.90
Sex organs	3.05	-	0.30	+	+
Eyes	0.11	-	0.03	+	+
Gall bladder, empty	0.08	-	0.03	+	-
Bile	0.39	-	0.11	+	-
Ears	-	-	0.08	-	-
Other waste	-	-	-	25.23	9.09

x In the case of stomach contents   + In the case of other waste

Source: (3) (8)

France

Table 10: Proportion of waste in absolute terms, based on average carcass weights shown in table 8

Units : kg

Type	Cattle	Calves	Pigs	Sheep	Horses
Hide	39.55	13.65	-	2.21	24.84
Rind	-	-	6.59	-	-
Bristles	-	-	0.43	-	-
Claws	1.51	-	0.05	-	+
Horns	0.40	-	-	-	-
Lungs	4.71	1.54	0.49	6.67	7.23
Spleen	0.96	0.42	0.13	0.12	1.69
Udder	2.75	-	-	0.55	-
Stomach, empty	18.16	2.25	0.87	1.65	2.28
Stomach contents	62.00	7.07	0.46	+	+
Trachea	0.96	0.65	0.30	+	+
Chitterlings	0.99	2.66	0.36	+	+
Blood	15.87	5.54	3.50	1.22	28.05
Intestines, empty	9.92	5.03	3.70	0.82	25.43
Intestine contents	18.60	x	2.87	-	56.05
Reticulum and small intestine	5.54	1.14	1.21	0.82	-
Bladder	0.31	-	0.87	-	-
Bones from slaughtered carcasses	52.57	22.78	9.99	3.07	61.30
From horn roots/head	11.43	4.00	-	-	-
From feet	4.83	-	-	0.61	21.41
Fat tissue	16.86	2.25	13.22	-	15.20
Sex organs	3.37	-	0.30	+	+
Eyes	0.12	-	0.03	+	+
Gall bladder, empty	0.09	-	0.03	+	-
Bile	0.43	-	0.11	+	-
Ears	-	-	0.08	-	-
Other waste	-	-	-	4.54	9.27

x In the case of stomach contents   + In the case of other waste

Source: (3) (8)

Italy

Table 11: Proportion of waste in absolute terms, based on average
carcass weights shown in table 8

Units : kg

Type	Cattle	Calves	Pigs	Sheep	Horses
Hide	31.38	14.22	-	0.98	12.94
Rind	-	-	6.97	-	-
Bristles	-	-	0.46	-	-
Claws	1.20	-	0.05	-	+
Horns	0.31	-	-	-	-
Lungs	3.73	1.61	0.52	0.30	3.76
Spleen	0.76	0.44	0.14	0.05	0.88
Udder	2.18	-	-	0.24	-
Stomach, empty	14.41	2.34	0.92	0.73	1.19
Stomach contents	49.20	7.37	0.48	+	+
Trachea	0.76	0.68	0.32	+	+
Chitterlings	0.78	2.78	0.38	+	+
Blood	12.59	5.78	3.70	0.54	14.61
Intestines, empty	7.87	5.24	3.95	0.36	13.25
Intestine contents	14.76	x	3.03	+	29.20
Reticulum and small intestine	4.40	1.18	1.28	0.36	-
Bladder	0.24	-	0.09	+	-
Bones from slaughtered carcasses	41.72	23.74	10.57	1.36	31.94
From horn roots/head	9.07	4.17	-	0.30	-
From feet	3.83	2.34	-	0.27	11.15
Fat tissue	13.38	-	13.98	-	7.92
Sex organs	2.68	-	0.32	+	+
Eyes	0.09	-	0.03	+	-
Gall bladder, empty	0.07	-	0.03	+	-
Bile	0.34	-	0.11	+	-
Ears	-	-	0.09	-	-
Other waste	-	-	-	2.01	4.83

x In the case of stomach contents   + In the case of other waste

Source: (3) (8)

Netherlands

Table 12: Proportion of waste in absolute terms, based on average carcass weights shown in table 8

Units : kg

Type	Cattle	Calves	Pigs	Sheep	Horses
Hide	35.34	15.08	–	3.07	21.46
Rind	–	–	6.29	–	–
Bristles	–	–	0.41	–	–
Claws	1.35	–	0.04	–	+
Horns	0.36	–	–	–	–
Lungs	4.21	1.71	0.07	0.93	6.24
Spleen	0.85	0.47	0.13	0.16	1.46
Udder	2.46	–	–	0.77	–
Stomach, empty	16.23	2.48	0.83	2.30	1.97
Stomach contents	55.40	7.82	0.43	+	+
Trachea	0.85	0.72	0.29	+	+
Chitterlings	0.88	2.95	0.34	+	+
Blood	14.18	6.13	3.34	1.69	24.24
Intestines, empty	8.86	5.56	3.56	1.14	21.97
Intestine contents	16.62	x	2.73	+	48.43
Reticulum and small intestine	4.95	1.26	1.16	1.14	–
Bladder	0.27	–	0.08	+	–
Bones from slaughtered carcasses	46.97	25.17	9.53	4.27	52.97
From horn roots/head	10.22	4.43	–	–	–
From feet	4.32	2.48	–	0.84	18.50
Fat tissue	15.06	–	12.61	–	13.13
Sex organs	3.01	–	0.29	+	+
Eyes	0.11	–	0.03	+	+
Gall bladder, empty	0.08	–	0.03	+	–
Bile	0.38	–	0.10	+	–
Ears	–	–	0.08	–	–
Other waste	–	–	–	6.30	8.01

x In the case of stomach contents   + In the case of other waste

Source: (3) (8)

Belgium

Table 13: Proportion of waste in absolute terms, based on average
carcass weights shown in table 8

Units   :   kg

Type	Cattle	Calves	Pigs	Sheep	Horses
Hide	38.79	13.50	-	2.94	26.21
Rind	-	-	6.06	-	-
Bristles	-	-	0.40	-	-
Claws	1.48	-	0.04	-	+
Horns	0.39	-	-	-	-
Lungs	4.62	1.53	0.45	0.90	7.62
Spleen	0.94	0.42	0.12	0.16	1.79
Udder	2.70	-	-	0.74	-
Stomach, empty	17.81	2.22	0.80	2.21	2.41
Stomach contents	60.80	7.00	0.42	+	+
Trachea	0.94	0.64	0.28	+	+
Chitterlings	0.97	2.64	0.33	+	+
Blood	15.58	5.48	3.22	1.62	29.60
Intestines, empty	9.72	4.98	3.44	1.09	26.82
Intestine contents	18.24	x	2.64	+	59.13
Reticulum and small intestine	5.44	1.12	1.12	1.09	-
Bladder	0.30	-	0.08	+	-
Bones from slaughtered carcasses	51.55	22.54	9.19	4.10	64.67
From horn roots/head	11.21	3.96	-	-	-
From feet	4.74	2.22	-	3.39	22.59
Fat tissue	16.53	-	12.16	-	16.03
Sex organs	3.31	-	0.28	+	+
Eyes	0.12	-	0.03	+	+
Gall bladder, empty	0.09	-	0.03	+	-
Bile	0.42	-	0.10	+	-
Ears	-	-	0.08	-	-
Other waste	-	-	-	25.23	9.78

x In the case of stomach contents   + In the case of other waste

Source: (3) (8)

Luxembourg

Table 14: Proportion of waste in absolute terms, based on average carcass weights shown in table 8

Units : kg

Type	Cattle	Calves	Pigs	Sheep	Horses
Hide	34.32	14.22	-	2.94	24.12
Rind	-	-	5.53	-	-
Bristles	-	-	0.36	-	-
Claws	1.31	-	0.04	-	+
Horns	0.34	-	-	-	-
Lungs	4.08	1.61	0.41	0.90	7.02
Spleen	0.83	0.44	0.11	0.16	1.65
Udder	2.39	-	-	0.74	-
Stomach, empty	15.76	2.34	0.73	2.21	2.22
Stomach contents	53.80	7.37	0.38	+	+
Trachea	0.86	0.68	0.25	+	+
Chitterlings	13.77	2.78	0.30	+	+
Blood	8.60	5.78	2.94	1.62	27.24
Intestines, empty	16.14	5.24	3.13	1.09	24.69
Intestine contents	4.81	x	2.40	+	54.42
Reticulum and small intestine	0.26	1.18	1.02	1.09	-
Bladder	45.62	-	0.07	+	-
Bones from slaughtered carcasses	0.78	23.74	8.38	4.10	59.52
From horn roots/head	9.92	-	-	-	-
From feet	4.19	2.34	-	3.39	20.79
Fat tissue	14.63	-	11.09	-	14.76
Sex organs	2.93	-	0.25	+	+
Eyes	0.10	-	0.02	+	+
Gall bladder, empty	0.08	-	0.02	+	-
Bile	0.37	-	0.09	+	-
Ears	-	-	0.07	-	-
Other waste	-	-	-	25.23	9.00

x In the case of stomach contents   + In the case of other waste

Source: (3) (8)

United Kingdom

Table 15: Proportion of waste in absolute terms, based on average
carcass weights shown in table 8

Unit : kg

Type	Cattle	Calves	Pigs	Sheep	Horses
Hide	32.79	4.88	-	2.45	21.70
Rind	-	-	4.85	-	-
Bristles	-	-	0.32	-	-
Claws	1.25	-	0.03	-	+
Horns	0.33	-	-	-	-
Lungs	3.90	0.55	0.36	0.75	6.31
Spleen	0.79	0.15	0.10	0.13	1.48
Udder	2.28	-	-	0.61	-
Stomach, empty	15.06	0.80	0.64	1.84	1.99
Stomach contents	51.40	2.53	0.33	+	+
Trachea	0.79	0.23	0.22	+	+
Chitterlings	0.82	0.95	0.26	+	+
Blood	13.15	1.98	2.57	1.35	24.51
Intestines, empty	8.22	1.80	2.75	0.91	22.22
Intestine contents	15.42	x	2.11	+	48.97
Reticulum and small intestine	4.60	0.40	0.89	0.91	-
Bladder	0.25	-	0.06	+	-
Bones from slaughtered carcasses	43.58	8.15	7.35	3.42	53.56
From horn roots/head	9.48	1.43	-	-	-
From feet	4.00	0.80	-	0.67	18.71
Fat tissue	13.98	-	9.72	-	13.28
Sex organs	2.80	-	0.22	+	+
Eyes	0.10	-	0.02	+	+
Gall bladder, empty	0.07	-	0.02	+	-
Bile	0.35	-	0.08	+	-
Ears	-	-	0.06	-	-
Other waste	-	-	-	5.04	8.1

x In the case of stomach contents   + In the case of other waste

Source: (3) (8)

Ireland

Table 16: Proportion of waste in absolute terms, based on average
carcass weights shown in table 8

Unit   :   kg

Type	Cattle	Calves	Pigs	Sheep	Horses
Hide	31.90	13.07	-	3.19	20.10
Rind	-	-	5.15	-	-
Bristles	-	-	0.34	-	-
Claws	1.22	-	0.04	-	-
Horns	0.32	-	-	-	+
Lungs	3.80	1.48	0.38	0.97	5.85
Spleen	1.72	0.40	0.10	0.17	1.37
Udder	2.22	-	-	0.80	-
Stomach, empty	14.65	2.15	0.68	2.39	1.85
Stomach contents	50.00	6.77	0.36	+	+
Trachea	0.77	0.62	0.23	+	+
Chitterlings	0.80	2.55	0.28	+	+
Blood	12.80	5.31	2.74	1.76	22.70
Intestines, empty	8.00	4.82	2.92	1.18	20.57
Intestine contents	15.00	x	2.24	+	45.35
Reticulum and small intestine	4.47	1.09	0.95	1.18	-
Bladder	0.25	-	0.06	+	-
Bones from slaughtered carcasses	42.40	21.82	7.81	4.44	49.60
From horn roots/head	9.22	3.84	-	1.0	-
From feet	3.90	2.15	-	0.88	17.32
Fat tissue	13.60	-	10.33	-	12.30
Sex organs	2.72	-	0.23	+	+
Eyes	0.10	-	0.02	+	+
Gall bladder, empty	0.07	-	0.02	+	+
Bile	0.35	-	0.08	+	-
Ears	-	-	0.06	-	-
Other waste	-	-	-	6.55	7.50

x In the case of stomach contents   + in the case of other waste

Source: (3) (8)

Denmark

Table 17: Proportion of waste in absolute terms, based on average
carcass weights shown in table 8

Units   :   kg

Type	Cattle	Calves	Pigs	Sheep	Horses
Hide	30.75	20.98	-	4.29	23.31
Rind	-	-	4.92	-	-
Bristles	-	-	0.32	-	-
Claws	1.18	-	0.03	-	+
Horns	0.31	-	-	-	-
Lungs	3.66	2.37	0.37	1.31	6.78
Spleen	0.74	0.65	0.10	0.23	1.59
Udder	2.14	-	-	2.00	-
Stomach, empty	14.12	3.46	0.65	3.22	2.14
Stomach contents	48.20	10.87	0.34	+	+
Trachea	0.74	1.00	0.22	+	+
Chitterlings	0.77	4.10	0.27	+	+
Blood	12.33	8.52	2.61	2.37	26.33
Intestines, empty	7.71	7.73	2.79	1.59	23.86
Intestine contents	14.46	x	2.14	+	52.60
Reticulum and small intestine	4.31	1.75	0.91	1.59	-
Bladder	0.24	-	0.06	+	-
Bones from slaughtered carcasses	40.87	35.01	7.46	5.98	57.53
From horn roots/head	8.89	6.16	-	1.34	-
From feet	3.75	-	-	1.18	20.09
Fat tissue	13.11	3.46	9.88	-	14.26
Sex organs	4.57	-	0.22	+	+
Eyes	0.09	-	0.02	+	+
Gall bladder, empty	0.07	-	0.02	+	-
Bile	0.33	-	0.08	+	-
Ears	-	-	0.06	-	-
Other waste	-	-	-	8.83	8.70

x In the case of stomach contents   + in the case of other waste

Source: (3) (8)

Tables 18 - 26 show the quantities of waste in thousands of tonnes by type of waste and animal.

The last column in each case gives the total quantity of the individual types.

The quantities determined in the above tables are mean values. They do not include the edible by-products marketed with the meat.

The following are edible by-products:

Heart

Liver

Tongue

Brains

Spinal cord

Kidneys

Prettitoes

Tail

Sweetbreads

Federal Republic of Germany

Table 18:   Incidence of waste in 1973

Units   :   '000 tonnes

Type	Cattle	Calves	Pigs	Sheep	Horses	Total
Hide	145.00	8.72	-	1.56	0.36	155.64
Rind	-	-	203.47	-	-	203.47
Bristles	-	-	13.42	-	+	13.42
Claws	5.56	-	1.52	-	-	7.08
Horns	1.46	-	-	-	-	1.46
Lungs	17.26	0.98	15.25	4.79	0.10	38.38
Spleen	3.49	0.26	4.27	0.85	0.02	8.89
Udder	10.11	-	-	3.94	-	14.05
Stomach, empty	66.61	1.43	26.84	11.77	0.03	106.68
Stomach contents	227.47	4.52	14.03	+	+	246.02
Trachea	3.49	0.41	9.15	+	+	13.05
Chitterlings	3.61	1.70	10.98	+	0.41	16.70
Blood	58.20	3.54	107.99	8.63	0.37	178.73
Intestines, empty	36.39	3.21	115.31	5.80	0.82	161.53
Intestine contents	68.24	x	88.46	+	-	156.70
Reticulum and small intestine	20.35	0.72	37.52	5.80	-	64.39
Bladder	1.13	-	2.44	+	_	3.57
Bones from slaughtered carcasses	190.91	14.56	308.41	21.85	0.91	536.64
From horn roots/head	41.96	2.56	-	-	-	44.52
From feet	17.71	-	-	18.06	0.31	36.08
Fat tissue	61.86	1.43	407.86	-	0.22	471.37
Sex organs	12.38	-	9.15	+	+	21.53
Eyes	0.44	-	0.91	+	+	1.35
Gall bladder, empty	0.32	-	0.91	+	+	1.23
Bile	1.58	-	3.35	+	-	4.93
Ears	-	-	2.44	-	-	2.44
Other waste	-	-	-	134.47	0.13	139.53

x In the case of stomach contents   + In the case of other waste

Source: (8)

France

Table 19:   Incidence of waste in 1973

Units   :   '000 tonnes

Type	Cattle	Calves	Pigs	Sheep	Horses	Total
Hide	140.87	43.17	-	16.00	3.72	203.76
Rind	-	-	115.62	-	-	115.62
Bristles	-	-	7.54	-	-	7.54
Claws	5.37	-	0.87	-	-	6.24
Horns	1.42	-	-	-	-	1.42
Lungs	16.77	4.87	8.59	48.29	1.08	79.60
Spleen	3.41	1.32	2.28	0.86	0.25	8.12
Udder	9.79	-	-	3.98	-	13.77
Stomach, empty	64.68	7.11	15.26	11.94	0.34	99.33
Stomach contents	220.84	22.36	8.07	+	+	251.27
Trachea	3.41	2.05	5.26	+	+	10.72
Chitterlings	3.52	8.41	6.31	+	+	18.24
Blood	56.52	17.52	61.41	8.83	4.20	148.48
Intestines, empty	35.33	15.90	64.92	5.93	3.80	125.88
Intestine contents	66.25	x	50.35	+	8.40	125.00
Reticulum and small intestine	19.73	3.60	21.23	5.93	-	50.49
Bladder	1.10	-	15.26	+	-	16.36
Bones from slaughtered carcasses	187.25	72.05	175.28	22.22	9.19	465.99
From horn roots/head	40.71	12.65	-	-	-	53.36
From feet	17.20	-	-	4.41	3.21	24.82
Fat tissue	60.05	7.11	231.95	-	2.28	301.39
Sex organs	12.00	-	5.26	+	+	17.26
Eyes	0.42	-	0.52	+	+	0.94
Gall bladder, empty	0.32	-	0.52	+	-	0.84
Bile	1.53	-	1.93	+	-	3.46
Ears	-	-	15.26	-	-	15.26
Other waste	-	-	-	32.87	1.39	34.26

x In the case of stomach contents   + In the case of other waste

Source: (8)

Italy

Table 20: Incidence of waste in 1973

Units : '000 tonnes

Type	Cattle	Calves	Pigs	Sheep	Horses	Total
Hide	121.67	15.07	-	6.15	4.12	147.01
Rind	-	-	51.16	-	-	51.16
Bristles	-	-	3.37	-	-	3.37
Claws	4.66	-	0.36	-	-	5.02
Horns	1.20	-	-	-	-	1.20
Lungs	14.50	1.70	3.81	1.88	1.19	23.08
Spleen	2.95	0.46	1.02	0.31	0.28	5.02
Udder	8.48	-	-	1.50	-	9.98
Stomach, empty	56.05	2.48	6.75	4.58	0.37	70.23
Stomach contents	191.38	7.81	3.52	+	+	202.71
Trachea	2.95	0.72	2.34	+	+	6.01
Chitterlings	3.03	2.94	2.78	+	+	8.75
Blood	48.97	6.12	27.16	3.39	4.66	90.30
Intestines, empty	30.61	5.55	28.99	2.26	4.22	71.63
Intestine contents	57.41	x	22.24	+	9.31	88.96
Reticulum and small intestine	17.11	1.25	9.39	2.26	-	30.01
Bladder	0.93	-	6.75	+	-	7.68
Bones from slaughtered carcasses	162.29	25.16	77.59	8.54	10.18	283.76
From horn roots/head	35.28	4.42	-	1.88	3.55	45.13
From feet	14.89	2.48	-	1.69	2.52	21.58
Fat tissue	52.04	-	102.62	-	+	154.66
Sex organs	10.42	-	2.34	+	+	12.76
Eyes	0.35	-	0.22	+	+	0.57
Gall bladder, empty	0.27	-	0.22	+	-	0.49
Bile	1.32	-	0.80	+	-	2.12
Ears	-	-	6.75	-	-	6.75
Other waste	-	-	-	12.62	1.54	14.16

x In the case of stomach contents   + In the case of other waste

Source: (8)

Netherlands

Table 21: Incidence of waste in 1973

Units : '000 tonnes

Type	Cattle	Calves	Pigs	Sheep	Horses	Total
Hide	24.95	14.50	-	1.19	0.17	40.81
Rind	-	-	60.98	-	-	60.98
Bristles	-	-	3.97	-	-	3.97
Claws	0.95	-	0.38	-	-	1.33
Horns	0.25	-	-	-	-	0.25
Lungs	2.97	1.64	0.67	0.36	0.04	5.68
Spleen	0.60	0.45	1.26	0.06	0.01	2.38
Udder	1.73	-	-	0.29	-	2.02
Stomach, empty	11.45	2.38	8.04	0.89	0.01	22.77
Stomach contents	39.11	7.52	4.16	+	+	50.79
Trachea	0.60	0.69	2.81	+	+	4.10
Chitterlings	0.62	2.83	3.29	+	+	6.74
Blood	10.01	5.89	32.38	0.65	0.19	49.12
Intestines, empty	6.25	5.34	34.51	0.44	0.17	46.71
Intestine contents	11.73	x	26.46	+	0.38	38.57
Reticulum and small intestine	3.49	1.21	11.24	0.44	-	16.38
Bladder	0.19	-	0.77	+	-	0.96
Bones from slaughtered carcasses	33.16	24.21	92.39	1.65	0.42	151.83
From horn roots/head	7.21	4.26	-	-	-	11.47
From feet	3.04	2.38	-	0.32	0.14	5.88
Fat tissue	10.63	-	122.25	-	0.10	132.98
Sex organs	2.12	-	2.81	-	-	4.93
Eyes	0.07	-	0.29	-	+	0.36
Gall bladder, empty	0.05	-	0.29	-	-	0.34
Bile	0.26	-	0.96	-	-	1.22
Ears	-	-	0.77	-	-	0.77
Other waste	-	-	-	2.44	0.06	2.50

x In the case of stomach contents   + In the case of other waste

Source: (8)

40

Belgium

Table 22:   Incidence of waste in 1973

Units   :   '000 tonnes

Type	Cattle	Calves	Pigs	Sheep	Horses	Total
Hide	29.17	3.07	–	0.39	0.41	33.04
Rind	–	–	43.54	–	–	43.54
Bristles	–	–	2.87	–	–	2.87
Claws	1.11	–	0.28	–	–	1.39
Horns	0.29	–	–	–	–	0.29
Lungs	3.47	0.34	3.23	0.12	0.12	7.28
Spleen	0.70	0.09	0.86	0.02	0.02	1.69
Udder	2.03	–	–	0.09	–	2.12
Stomach, empty	13.39	0.50	5.74	0.30	0.03	19.96
Stomach contents	45.72	1.59	3.01	+	+	50.32
Trachea	0.70	0.14	2.01	+	+	2.85
Chitterlings	0.72	0.60	2.37	+	+	3.69
Blood	11.71	1.24	23.13	0.22	0.47	36.77
Intestines, empty	7.30	1.13	24.71	0.14	0.42	33.70
Intestine contents	13.70	x	18.97	+	0.94	33.61
Reticulum and small intestine	4.09	0.25	8.04	0.14	–	12.52
Bladder	0.22	–	0.57	–	–	0.79
Bones from slaughtered carcasses	38.76	5.13	66.03	0.55	1.03	111.50
From horn roots/head	8.42	0.90	–	–	–	9.32
From feet	3.56	0.50	–	0.11	0.36	4.53
Fat tissue	12.43	–	87.38	–	0.25	100.06
Sex organs	2.48	–	2.01	+	–	4.49
Eyes	0.09	–	0.21	+	–	0.30
Gall bladder, empty	0.06	–	0.21	+	–	0.27
Bile	0.31	–	0.71	+	–	1.02
Ears	–	–	0.57	–	–	0.57
Other waste	–	–	–	0.82	0.15	0.97

x In the case of stomach contents   + in the case of other waste

Source: (8)

Luxembourg

Table 23: Incidence of waste in 1973

Units : '000 tonnes

Type	Cattle	Calves	Pigs	Sheep*	Horses*	Total
Hide	0.99	0.01	-			1.00
Rind	-	-	0.68			0.68
Bristles	-	-	0.04			0.04
Claws	0.03	-	0.00			0.03
Horns	0.00	-	-			-
Lungs	0.11	0.00	0.05			0.16
Spleen	0.02	0.00	0.01			0.03
Udder	0.06	-	-			0.06
Stomach, empty	0.45	0.00	0.08			0.53
Stomach contents	1.56	0.00	0.04			1.60
Trachea	0.02	0.00	0.03			0.05
Chitterlings	0.02	0.00	0.03			0.05
Blood	0.39	0.00	0.36			0.75
Intestines, empty	0.24	0.00	0.38			0.62
Intestine contents	0.46	x	0.29			0.75
Reticulum and small intestine	0.13	0.00	0.12			0.25
Bladder	0.00	-	0.00			-
Bones from slaughtered carcasses	1.32	0.02	1.03			2.37
From horn roots/head	0.28	-	-			0.28
From feet	0.12	0.00	-			0.12
Fat tissue	0.42	-	1.32			1.74
Sex organs	0.08	-	0.03			0.11
Eyes	0.00	-	0.00			-
Gall bladder, empty	0.00	-	0.00			-
Bile	0.10	-	0.01			0.11
Ears	-	-	0.08			0.08
Other waste	-	-	-			-

x in the case of stomach contents   + in the case of other waste

Source: (8)   * Insignificant

United Kingdom

Table 24: Incidence of waste in 1973

Units : '000 tonnes

Type	Cattle	Calves	Pigs	Sheep	Horses*	Total
Hide	108.43	0.69	-	28.92		138.04
Rind	-	-	73.37	-		73.37
Bristles	-	-	4.84	-		4.84
Claws	4.13	-	0.45	-		4.58
Horns	1.09	-	-	-		1.09
Lungs	12.89	0.07	5.44	8.85		27.25
Spleen	2.61	0.02	1.52	1.53		5.68
Udder	7.53	-	-	7.20		14.73
Stomach, empty	49.80	0.11	9.68	21.72		81.31
Stomach contents	169.97	0.35	4.99	+		175.31
Trachea	2.61	0.03	3.32	+		5.96
Chitterlings	2.71	0.13	3.93	+		6.77
Blood	43.48	0.28	38.87	15.93		98.56
Intestines, empty	27.18	0.25	41.60	10.74		79.77
Intestine contents	50.99	x	31.92	+		82.91
Reticulum and small intestine	15.21	0.05	13.46	10.74		39.46
Bladder	0.81	-	0.90	+		1.71
Bones from slaughtered carcasses	144.11	1.15	111.19	40.37		296.82
From horn roots/head	31.35	0.20	-	-		31.55
From feet	13.22	0.11	-	7.90		21.23
Fat tissue	46.23	-	147.04	-		193.27
Sex organs	9.25	-	3.32	+		12.57
Eyes	0.33	-	0.30	+		0.63
Gall bladder, empty	0.23	-	0.30	+		0.53
Bile	1.15	-	1.21	+		2.36
Ears	-	-	0.90	-		0.90
Other waste	-	-	-	59.49		59.49

x In the case of stomach contents   + In the case of other waste

Source: (8)   * Insignificant

Ireland

Table 25: Incidence of waste in 1973

Units   :   '000 tonnes

Type	Cattle	Calves	Pigs	Sheep	Horses	Total
Hide	27.62	0.05	-	5.39	0.16	33.22
Rind	-	-	10.87	-	-	10.87
Bristles	-	-	0.71	-	-	0.71
Claws	1.05	-	0.08	-	-	1.13
Horns	0.27	-	-	-	-	0.27
Lungs	3.29	0.00	0.80	1.63	0.04	5.76
Spleen	1.48	0.00	0.21	0.28	0.01	1.98
Udder	1.97	-	-	1.35	-	3.32
Stomach, empty	12.68	0.00	1.43	4.03	0.01	18.15
Stomach contents	43.30	0.02	0.75	+	+	44.07
Trachea	0.66	0.00	0.48	+	+	1.14
Chitterlings	0.69	0.01	0.59	+	+	1.29
Blood	11.08	0.02	5.78	2.97	0.18	20.03
Intestines, empty	6.92	0.01	6.16	1.99	0.16	15.24
Intestine contents	12.99	x	4.72	+	0.36	18.07
Reticulum and small intestine	3.87	0.00	2.00	1.99	-	7.86
Bladder	0.21	-	0.12	+	-	0.33
Bones from slaughtered carcasses	36.71	0.08	16.48	7.50	0.39	61.16
From horn roots/head	7.98	0.01	-	1.69	-	9.68
From feet	3.37	0.00	-	1.48	0.13	4.98
Fat tissue	11.77	-	21.80	-	0.09	33.66
Sex organs	2.35	-	0.48	+	+	2.83
Eyes	0.08	-	0.04	+	+	0.12
Gall bladder, empty	0.06	-	0.04	+	-	0.10
Bile	0.30	-	0.16	+	-	0.46
Ears	-	-	0.21	-	-	0.21
Other waste	-	-	-	11.06	0.06	11.12

x In the case of stomach contents   + In the case of other waste

Source: (8)

44

Denmark

Table 26: Incidence of waste in 1973

Units : '000 tonnes

Type	Cattle	Calves	Pigs	Sheep	Horses	Total
Hide	13.03	9.33	-	0.00	0.58	22.94
Rind	-	-	56.20	-	-	56.20
Bristles	-	-	3.65	-	-	3.65
Claws	0.50	-	0.34	-	-	0.84
Horns	0.13	-	-	-	-	0.13
Lungs	1.55	1.05	4.22	0.00	0.16	6.98
Spleen	0.31	0.28	1.14	0.00	0.03	1.76
Udder	0.90	-	-	0.00	-	0.90
Stomach, empty	5.98	1.53	7.42	0.00	0.05	14.98
Stomach contents	20.43	4.83	3.88	+	+	29.14
Trachea	0.31	0.44	2.51	+	+	3.26
Chitterlings	0.32	1.82	3.08	+	+	5.22
Blood	5.22	3.79	29.81	0.00	0.65	39.47
Intestines, empty	3.26	3.43	31.87	0.00	0.59	39.15
Intestine contents	6.13	x	24.44	+	1.31	31.88
Reticulum and small intestine	1.82	0.77	10.39	0.00	-	12.98
Bladder	0.10	-	0.68	-	-	0.78
Bones from slaughtered carcasses	17.32	15.57	85.22	0.01	1.43	119.55
From horn roots/head	3.76	2.74	-	0.00	-	6.50
From feet	1.59	-	-	0.00	0.50	2.09
Fat tissue	5.55	1.53	112.86	-	0.35	120.29
Sex organs	1.93	-	2.51	+	+	4.44
Eyes	0.03	-	0.22	+	+	0.25
Gall bladder, empty	0.02	-	0.22	+	-	0.24
Bile	0.13	-	0.91	+	-	1.04
Ears	-	-	0.68	-	-	0.68
Other waste	-	-	-	0.01	0.21	0.22

x In the case of stomach contents   + In the case of other waste

Source: (8)

The lungs frequently come under the heading of slaughtering waste, being little used in human nutrition because of consumption habits. They are generally condemned in the Scandinavian countries. The lungs are also often disregarded during the actual slaughtering operation, eg the lungs of the pig fill with water on slaughtering and are thus unusable for human consumption.

Some 10 per cent of the blood of the pig is used for human nutrition. The blood flows into the sewers in small to medium sized abattoirs and older slaughterhouses.

The bones of the carcass do not occur as waste during the actual slaughtering process but only when the carcass is cut up. This is usually carried out at butchers or meat cutters (meat processors). It is possible to determine the quantities of bones in cases where slaughtered animals are cut up in large quantities.

Some of the bones are purchased with the meat. It is impossible to determine and evaluate these quantities because the collection of these details would not be economic.

5.1 TYPES AND QUANTITIES OF POULTRY WASTE

Total poultry waste is about 20 per cent of the live weight[7]. It is broken down as follows:

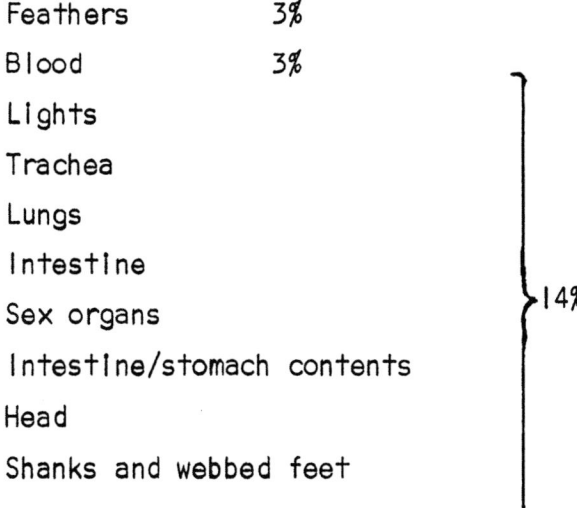

Feathers	3%
Blood	3%
Lights	
Trachea	
Lungs	
Intestine	
Sex organs	14%
Intestine/stomach contents	
Head	
Shanks and webbed feet	

Table 27 gives the waste broken down by country and type.

Table 27: Incidence of waste by type and country for 1973

Units : tonnes

Type	FRG	France	Italy	Netherlands	Belgium	Luxem-bourg	UK	Ireland	Denmark
Feathers	10,987	29,662	30,225	12,712	4,350	37	24,862	1,612	3,375
Blood	10,987	29,662	30,225	12,712	4,350	37	24,862	1,612	3,375
Other waste	51,275	138,425	141,050	59,325	20,300	175	116,025	7,525	15,750

Source: (8)

6 Analysis of present disposal methods

As already stated, slaughterhouse waste is disposed of by the following methods:

1. Waste water
2. Special utilization
3. Carcass disposal institutions
4. Burying
5. Direct utilization in agriculture
6. Biogas plants

All the disposal methods listed are used in the EEC (see figure 1).

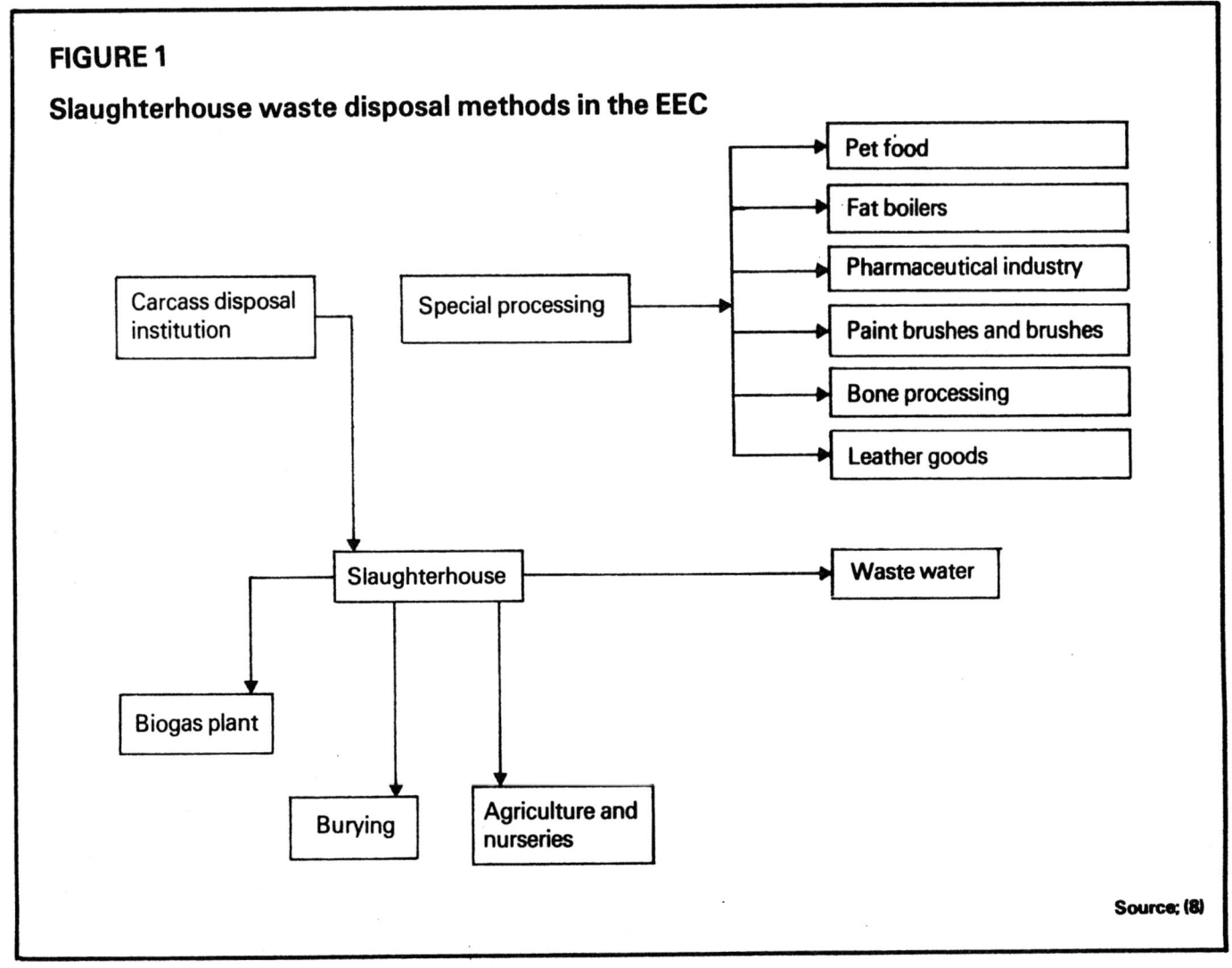

FIGURE 1

Slaughterhouse waste disposal methods in the EEC

Source: (8)

1. *Waste water*

The disposal of slaughterhouse waste by way of waste water can be carried out to only a limited degree; the following are mainly disposed of via waste water:

Blood, bristles, urine, faeces, bone splinters, meat residues, fat, stomach and intestine contents.

2. *Special utilisation*

This term covers the utilisation of slaughterhouse waste in certain branches of industry. The various kinds of waste are collected at the slaughterhouse and collected by processors from the slaughterhouse.

The petfood industry is taking an increasing part in the utilisation of slaughterhouse waste. However, this branch of industry can only utilise waste which has not been condemned, eg the rumen, stomach, and chitterlings. Catfood, dogfood and fishfood are mainly made from these.

Fat boilers take the waste fat and process it to semi-products used mainly in the detergent industry.

Glands are used by the pharmaceutical industry for the production of drugs.

The brush industry requires bristles and hair.

The bone-processing industry processes some 50 per cent of the bones to form gelatin and glue.

Skins and hides are used for leather in the leather goods industry.

3. *Carcass disposal institutions*

Carcass disposal institutions process the bones (about 20 per cent) and condemned meat, fat and carcasses to produce meal and industrial fats. Feather meal, dried blood and poultry residue meal are made from poultry slaughterhouse waste at the carcass disposal institutions.

There are certain alternatives among the undertakings carrying out special processing and the animal carcass disposal institutions, ie slaughterhouses killing only small numbers of animals are usually unable to provide hygienically acceptable storage of the waste for special utilisation. The transport of unduly small quantities is also uneconomic. Since special processors utilise the waste only from purely economic aspects, any waste which is not processed by other branches of industry has to be disposed of harmlessly by the carcass disposal institutions.

Within the terms of reference of this study, the utilisation of slaughterhouse waste in carcass disposal institutions will be given priority treatment.

Table 5 lists the quantities that have to be processed by the carcass disposal institutions in the various EEC Member States. For the reasons indicated above, there may be changes which affect the carcass disposal institutions as compared with the special processors.

Some 25 per cent of the bones (from carcasses, feet and heads) go to the customer with the meat and are thus lost to the processing trade because they are collected with domestic refuse.

20 per cent of the bones go to the carcass disposal institutions, the rest being collected for special bone processors.

The carcass processors process some 90 per cent of the slaughterhouse waste and 10 per cent of dead animals and animals which have had to be killed because of disease without the blood being withdrawn. The

50

figures are as follows:

    0.5% of the cattle stock

    10%   of the stock of calves

    2% of the pig stock[9]

These figures relate to the Federal Republic of Germany and may differ somewhat in the other EEC Member States.

The most usual disposal methods are:

1. Dry process
2. Perchloroethylene process (also known as the azeotropic process)
3. Wet process.

1: Dry process

In the dry process, the raw material is placed in a boiler, heated and sterilised. The boiler is frequently preceded by a pre-boiler used to sterilise and break down the material. The main boiler is then required only for drying purposes. The sterilisation and disintegration process takes some 30 to 50 minutes. The boiler is then depressurised and the drying stage is started. In this stage the water is removed to a residual content of 8 - 10 per cent, the operating time being about one to four hours. The fat is removed from the meat pulp by means of hydraulic presses. The end product is meat meal and industrial fats.

2: Perchloroethylene process

In this process, the raw material is comminuted, sterilised and disintegrated as in the dry process. A solvent (perchloroethylene) is used to remove the fat from the meat pulp in various stages. With this process the resulting meat meal contains only a minor proportion of fat and is practically odourless. The meal has to be ground because of its granular structure.

3: *Wet process*

Unlike the dry process, direct steam is used for sterilisation and disintegration. The disadvantage of this process is that material which has been sterilised and disintegrated with direct steam is additionally enriched in water vapour which has to be removed during the subsequent drying stage. The products: meat meal, fat and glue liquor, are also of a lower grade, because the meat meal has a relatively high fat content and the fat has a relatively high glue content.

4: *Burying*

The burying of slaughterhouse waste is becoming increasingly insignificant in Europe because it is the poorest method for hygienic reasons. If the waste is not dumped properly there is a risk of contamination of groundwater.

5: *Direct utilisation in agriculture*

Direct utilisation of slaughterhouse waste in agriculture is:
1. Possible - by way of direct feeding, mainly to fattening pigs.
2. The utilisation of the dung as organic fertiliser.

6: *Biogas plants*

Slaughterhouse waste can be used in biogas plants to produce methane gas by a digestion process. If there is a sufficient quantity of slaughterhouse waste, the gas produced can be utilised as an energy source.

6.1 EVALUATION OF THE INDIVIDUAL DISPOSAL METHODS FROM ECONOMIC AND ENVIRONMENTAL ASPECTS

1. *Waste water*

   a) Economic aspects

   The disposal of slaughterhouse waste via the waste water system
   is economic to the slaughterhouse if no clarification of the
   waste water is required before it enters the communal waste
   water system. If pre-clarification is specified, the slaughterhouse will ensure - for cost reasons - that only small quantities of waste enter the waste water system.

   b) Environmental aspects

   From pollution considerations it is unacceptable that slaughterhouse waste water should be introduced into a public sewer
   system without prior cleansing.

2. *Special utilisation*

   a) Economic aspects

   The utilisation of slaughterhouse waste in special undertakings
   is carried out basically from economic considerations. If
   there is no economic advantage, then processing will not be
   carried out in the special undertakings, ie the raw material
   will be left to the carcass disposal institutions to dispose
   of harmlessly.

   b) Environmental aspects

   The utilisation of slaughterhouse waste in special undertakings
   is compatible with the environment if transport and processing
   are carried out in closed systems and do not come into contact
   with the environment.

3. *Carcass disposal institutions*

   a) Economic aspects

   The economic side of carcass disposal institutions frequently
   depends on transport costs and the quantities of raw materials

to be processed. The function of the carcass disposal institutions is not solely an economic one, because under their
terms of reference they are obliged to dispose harmlessly of
all problematic waste in order to maintain public health.

b) Environmental aspects

The transport, storage and processing of slaughterhouse waste
must be carried out in closed systems in order to avoid pollution. Pathogens are substantially eliminated from the waste
water, but for odour reasons it is necessary for the waste ·
water to be subjected to special treatment before being introduced into the public sewer system.

4. *Burying*

a) Economic aspects

The burying of slaughterhouse waste may be considered an
advantageous disposal method economically for small and medium-
sized slaughterhouses.

b) Environmental aspects

The disposal of slaughterhouse waste by burying in controlled
and uncontrolled dumps is not recommended, for reasons associated with pollution and the risk of disease. With present-
day population density, contamination of the human environment
by problematic waste may have unforeseen consequences.

5. *Direct utilisation in agriculture*

a) Economic aspects

The use of slaughterhouse waste as fodder is quite acceptable
economically, provided that no other economic utilisation,
eg special utilisation, is possible. The quantities of manure
can be economically utilised in rural districts as an organic
manure in agriculture. In urban districts the economy is
greatly restricted by the high transport costs. If a municipal composting plant is available, these quantities of manure
can be more economically included in the composting process.

b) Environmental aspects

The direct feeding of slaughterhouse waste to animals, particularly fattening pigs, is not recommended on the grounds of hygiene and maintaining human and animal health. In many countries, the direct use of slaughterhouse waste as animal feed is prohibited by law.

The disposal of the quantities of manure occurring at the slaughterhouse can be carried out by agriculture in rural districts. In urban districts, on the other hand, the disposal of the quantities of manure occurring will cause considerable problems for slaughterhouses. Because of the high costs of transport there are frequently no customers. In these areas it would be logical and compatible with the environment for the waste to be disposed of by way of a communal composting plant.

6. *Biogas plants*

a) Economic aspects

A biogas plant is economically worthwhile in small and medium-sized undertakings which are unfavourably situated in relation to other utilisation undertakings.

A plant of this kind requires a continuous supply of waste, but since waste does not occur continuously at most medium-sized and small slaughterhouses, this type of utilisation has not become established.

b) Environmental aspects

This type of waste disposal can be regarde as compatible with the environment because it does not cause any pollution.

6.2   DISPOSAL METHODS IN THE INDIVIDUAL EEC MEMBER STATES

Federal Republic of Germany

At the present time (1974) there are 106 carcass disposal institutions
in the Federal Republic of Germany (including West Berlin). The rural
districts and independent towns have combined to overcome the problem
or form larger catchment areas. Each institution (TBA) has officially
specified catchment areas in which it is responsible for the harmless
disposal of killed and stillborn solipeds such as cattle, pigs, sheep
and goats, where they are not used for human nutrition. The basis for
this is the Carcass Disposal Act[10] and its Implementation Orders[11].

The administrative authorities erect disposal plants of this kind and
transfer the responsibility partly to lessees or private undertakings.

France

The French authorities have not given any information as to legislation
for the disposal of slaughterhouse waste, its organisation, or the
disposal methods.

Italy

According to an investigation by Vanselow[3], there was one processing
undertaking in Italy from 1968 to 1970. It may be assumed that a very
considerable amount of slaughterhouse waste is not further processed
or utilised in Italy. The authorities have not given any further
information.

Netherlands

Four private and one communal disposal institution receive the waste
material in the Netherlands and process it to form a semi-product.
These semi-products are further processed by a special institution.

56

Belgium

In Belgium, slaughterhouse waste is disposed of by carcass disposal undertakings which operate on economic considerations and are responsible for a regional area. All slaughterhouse waste is dealt with and utilised.

Luxembourg

In Luxembourg, slaughterhouse waste is dealt with and processed through the agency of a national knackery.

United Kingdom

Waste is disposed of by private undertakings which process the waste only from economic aspects.

The private undertakings are inspected by the local authorities to ensure compliance with statutory regulations. Whole carcasses are frequently buried.

Ireland

No information.

Denmark

The processing of slaughterhouse waste in Denmark is carried out partly by the State and partly by private enterprise. All slaughterhouse waste is dealt with and processed by these undertakings.

7 Analysis of existing or proposed statutory regulations for the disposal of slaughterhouse waste in the EEC Member States

FEDERAL REPUBLIC OF GERMANY

a) *Existing:*

- Carcass Disposal Act of I February 1939 recently amended by the Act to Facilitate Administrative Reform in the Länder.
- Devolution Act of 10 March 1975
- The Carcass Disposal Act Implementation Order of 23 February 1939, amended by
- Devolution Order of 18 April 1975
- Carcass Disposal Act Implementation Order of 17 April 1939
- Livestock Diseases Act of 1909
- Meat Inspection Act of 3 June 1900
- Poultry Meat Hygiene Act of 12 July 1973
- Waste Disposal Act of 7 June 1972

b) *Planned:*

A Bill for a new version of the Carcass Disposal Act was presented on II February 1975.

The new Carcass Disposal Act came into force on 2 September 1975, cancelling the above Carcass Disposal Act of 1939 and the Implementation Orders.

FRANCE

a) *Existing:*

- Classified Companies Act of 19 December 1917
- Circular of 6 June 1953 for implementation of this Act
- Decision of the Ministry of Agriculture dated II April 1967
- Act 'In the Interest of Promoting Public Health' with Articles 258 to 275.

58

b) Planned:

According to Information from the Ministry, there are at the present time no planned new versions or amendments of these Acts.

ITALY

a) Existing:

There are at the present time various regulations concerned with the disposal and processing of slaughterhouse waste. These regulations are found In the:
- *Regolamento Nazionale Solla Vigilanza Sanitaria delle Carni* and
- *Regolamento di Polizia Vetereniraria*

b) Planned:

The current regulations have proved appropriate according to the Ministry of Health. For this reason, no new measures have been presented to Parliament at the present time.

NETHERLANDS

a) Existing:
- Destruction Act

b) Planned:
- Amendment of the Destruction Act in respect of the processing of slaughterhouse waste.

BELGIUM

a) Existing:
- The Regent's Decree of 24 January 1946 in respect of the disposal of animals unfit for human consumption.
- Completed by the Royal Orders of 3 August 1955 and 25 July 1960
- Amended by the Royal Orders of 21 March 1964 and 15 April 1965

- Act of 5 September 1952 concerning Inspection of the Meat Trade
- Royal Decree of 21 September 1952 in respect of Inspection of Poultry Meat Trade
- Royal Decree of 20 September 1958, which contains Veterinary Inspectors' standards in respect of the import, preparation, production, transport and trade in bones, meal and other products originating from animals, this Decree having been amended by the Royal Decree of 26 June 1959
- Knackery Agreement of 1 January 1968
- Royal Decree of 24 January 1969, which contains Health Inspectors' Standards in respect of sludge irrigation fields and refuse sludge irrigation fields, and in respect of the use of organic waste and kitchen waste for the feeding of domestic animals (Decree of 23 April 1969)
- Surface Water Protection Act of 26 March 1971
- Royal Decree of 23 January 1974 concerning general conditions for discharging waste water into public waste water sewers.

b) *Planned:*

The Ministries approached could not give any clear information concerning proposed new Acts and amendments in respect of Acts concerning the processing of slaughterhouse waste.

LUXEMBOURG

a) *Existing:*
- Veterinary Inspectors' Act of 29 June 1912
- Grand Duchy Decree of 7 June 1948 concerning Implementation of the Act of 29 June 1912
- Act governing the Destruction and Utilisation of Carcasses (cheap meat and meat waste) of 6 September 1962
- Grand Duchy Order of 27 March 1973 concerning the Coming into Force of Articles 2, 3, 4 and 6 of the Act of 6 September 1962 concerning the Destruction and Utilisation of Carcasses (cheap meat and meat waste). This Order cancels Articles 50 to 63 inclusive of the Grand Duchy Decree of 7 June 1948 relating to

the implementation of the Veterinary Inspectors' Act of 29 June
1912, as at 1 April 1973.

- Grand Duchy Order of 4 July 1973 stipulating the Implementation of the Act of 6 September 1962 concerning the Destruction
and Utilisation of Animal Carcasses (cheap meat and meat waste).

b) Planned:

The Ministry of Agriculture were unable to give any information
in respect of planned legislation.

UNITED KINGDOM

There is no trade legislation concerning the use of waste in general,
but the following legislation tries to make regulations in this matter.

a) Existing:
- Public Health Act of 1936
- Alkali Act of 1966. This Order controls certain preparations
used in the processing of by-products.
- Dogs Act. The Dogs Act prohibits the display of carcasses or
waste generally.
- Slaughterhouse Order of 1958
- Meat Sterilisation Order
- 1972 Toxic Waste Dumping Order
- 1973 Animal Diseases Regulations
- 1974 Pollution Control Act

b) Planned:

Draft protein processing regulations. These are intended to
control the processing of animal proteins.

IRELAND

Letters were written to the Embassies and a number of Ministries, and
reminders were sent a number of times for answers to the questions,
in order to obtain corresponding basic material for Ireland, but no

Information on Ireland was available by the date of preparation of
the final report.

DENMARK

a) *Existing:*
- Animal Utilisation Act No. 424 of 30 June 1922
- Ministry of Agriculture Circular of I July 1939 relating to
  the Transportation of Slaughterhouse Waste
- Animals Order of I December 1959
- Ministry of Agriculture Circular of 17 December 1960 concerning
  the Delivery of Meat and Slaughterhouse Waste to Fodder
  Factories
- Act No. 220 of 26 April 1973 in respect of Meat Found Unfit
  for Human Consumption
- Act Regulations No. 482, Part 15, of 27 September 1974,
  concerning Meat for the Home Market
- Act Regulations No. 190, Chapter 6, of 22 May 1975 in con-
  nexion with Meat Export.

b) *Planned:*
At the present time there are no proposed improvements to the
above legislation.

The EEC Directive 'Fresh Meat' of 28 June 1965 and the Amendment Acts
that have been carried out apply to all the EEC Member States.

8 Organisation of the disposal of slaughterhouse waste in the EEC Member States

FEDERAL REPUBLIC OF GERMANY

In the Federal Republic of Germany no waste may be buried. All waste
unfit for human consumption must be disposed of harmlessly.

The transport of slaughterhouse waste beyond State boundaries, to be
buried there, is not allowed.

The carcass disposal institutions with their allotted catchment areas
are obliged to dispose harmlessly of the slaughterhouse waste, including carcasses, occurring in that area. In many cases, the raw material
is made available to the carcass disposal institutions without charge.
Only the transport costs are incurred. Those processing undertakings
which can be classed as special processing undertakings, eg the
chemical industry, pharmaceutical industry, and so on, are frequently
prepared to pay for their raw materials in the Federal Republic of
Germany.

Since the individual carcass disposal institutions frequently incur
considerable transport and processing costs, they are often unable
to work economically.

The amount of subsidies varies and is usually borne by towns and
districts; however, subsidies are granted only for those undertakings
which cannot operate profitably because of the structural position.

The end products (animal meat, industrial fat, etc) are sold on the
open market according to supply and demand.

The use of fodder and the sale of these products is governed by the
Fodder Act of 22 December 1926 and the Implementation Orders.

FRANCE

Slaughterhouse waste and carcasses are collected and processed by knackers in France, a small number of slaughterhouses carrying out this processing themselves.

Slaughterhouse waste water is either treated at a specific place at the slaughterhouse or discharged into the communal sewage plant or directly into rivers, although this is relatively rare according to information from the Ministry.

The Ministry of the Environment recommends that waste water from slaughterhouses and knackers should be treated.

There are 130 knackers (carcass disposal institutions) in France responsible for processing and disposing of animal carcasses and slaughterhouse waste. The capacity of these 130 knackers is, however, insufficient to process the total quantity. Not all districts are covered or visited by the knackers, since the distances are frequently too great. The law provides that those communities which are not covered by knackers can bury such waste if they comply with certain precautions. The community is obliged to set aside land situated outside the individual or collective population zones and at a distance of at least 100 metres from the population and also at least 100 metres from a well or water source or water courses. This land must be enclosed by fences to prevent access by other animals. No fodder may be grown on this land. Plants growing on this land must be burnt. A Bill on this problem has been presented to Parliament for discussion and an amendment has been proposed.

Any transport of slaughterhouse waste beyond national frontiers is prohibited. The carcass disposal institutions are subject to the Slaughterhouse Inspectorate, and are controlled by the latter. The knackers frequently receive disposal compensation from slaughterhouses for the disposal of slaughterhouse waste, and from the communes for the disposal of carcasses, to cover transport costs. In 1967 the

communes paid an average of FF 0.30 for each slaughtered animal.
Slaughterhouses are increasingly adopting the system of paying lump
sums for the disposal of all the waste. The amounts were FF 120 to
FF 200 per annum in 1967. A payment is also frequently made in proportion to the tonnage slaughtered. This is based on FF 4 per tonne.

According to information from the Ministry of the Environment, a small
number of disposal institutions use the slaughterhouse waste for breed-
ing maggots or preparing compost. Most disposal institutions have a
sterilisation plant for re-processing the waste to form high-grade
fodder. The wet process is also being increasingly relegated to the
background in France and is being displaced by new processes (the dry
process and the perchloroethylene process). The sale of the industrial
fat produced is carried out through middlemen in the form of a credit
sale. On the other hand, the fodder meals produced by the knackers
are frequently sold on a contract basis to fodder manufacturers.

ITALY

In Italy, waste unfit for human consumption and which cannot be processed further for industrial purposes (special utilisation by technological incineration equipment or other appropriate methods - various
processing methods) is disposed of under public health inspectorate
supervision. The relevant regulations are contained in the Acts
listed. Depending upon the processing and sterilisation of the waste
and carcasses, the end products (meal and industrial fat) may be used
further in the relevant branches of industry.

All processing methods relating to slaughterhouse waste and carcasses
are subject to public health inspection and must be carried out under
environmental and economic considerations.

NETHERLANDS

In the Netherlands, the obligation as to disposal of waste lies with
the State. The State has transferred the obligation as to disposal

66

to private institutions. There are four private destructors in the
Netherlands, one of which has a limited area. There is also a municipal destructor in the Netherlands processing the material of two
municipalities to form semi-products. These semi-products are taken
by the other large destructors and processed further into industrial
fat and meal.

In the case of material condemned as unfit, the processing is stipu-
lated to the destructor under the Destruction Act. The regulations
specify a heating process in which the material is heated for at least
four hours at 105⁰ and above. The destruction undertaking is directed
to collect all slaughterhouse waste and animal carcasses for its area
and subject them to a sterilisation process. A revision of the
Destruction Act is in preparation in the Netherlands. It is intended
to include in the new Destruction Act waste which at present is subject to other regulations.

BELGIUM

There are two industrial networks in Belgium instructed by the State
to collect slaughterhouse waste and animal carcasses. The first
includes those industries which utilise slaughterhouse waste by special
processes, ie the leather goods industry, the gelatine industry, the
glue industry, the pharmaceutical industry, etc. These special branches
of industry collect and process such slaughterhouse waste as has not
been condemned.

The disposal of animal carcasses and of slaughterhouse waste which is
condemned and is not processed by the above industries, is dealt with
by the carcass disposal institutions. There are four such institutions in Belgium, each responsible for a precisely allocated region.
These carcass disposal institutions have been instructed by the Ministry of Agriculture to collect and dispose harmlessly of carcasses and
parts thereof. They are private institutions which are supervised by
the Ministry of Public Health. The specifications for the sterilisation treatment stipulate a heat of 140°C for a period of 180 minutes.

The transport of slaughterhouse waste and carcasses may be carried out only by knackers or licensed institutions. These knackers are not obliged to pay for the waste for collection, but transport is at their expense.

The Decree of 24 January 1946 concerning the disposal of dead animals unfit for consumption allows dogs, cats and other small animals to be buried.

Under the Ministerial Decree of 28 July 1971 in connexion with the import, export and transit of live animals and certain vegetable and animal products, the import of waste is permitted from the Benelux countries, France, the Federal Republic of Germany, Italy, the UK, Ireland, Denmark, Norway and Sweden. The export of waste is also allowed, in which case the importing country makes the conditions.

The animal disposal institutions receive no state subsidies; they must therefore be economic. The products from these carcass disposal institutions are sold on the open market. The sale of these products is governed by the Royal Order of 20 September 1958 and 26 June 1959 concerning the determination of veterinary examination in respect of imports, fabrication, transport and trade in bones, meal and other animal products. A licence is required for the import of these products. These products must also be labelled clearly showing that they are animal products. The use of these products as a fodder in respect of the qualified composition is controlled by the Royal Order of 12 July 1972.

LUXEMBOURG

In Luxembourg, all slaughterhouse waste is processed by a State knackery (carcass disposal institution). This is controlled by the Luxemburgische Bauernzentrale. The collection of condemned meat and waste is done in special containers. These special containers must be collected at least once a week, the costs of transport, however, being borne by the processor. All slaughterhouse waste and carcasses are processed economically, and the material must be subjected to sterilisation at 140°C for at least several hours.

In the event of the risk of imminent infection, carcasses of less than 25kg weight, and animals which have died of certain infectious diseases, may exceptionally be buried on private plots under the official veterinary regulations.

The State processing institute has a modern biological clarification plant for purification of the waste water. The latter is subjected to sterilisation before being discharged into the clarification plant.

Under the Act of 6 September 1962, it is prohibited to bury or destroy cheap meat and meat waste outside the central knackery. The establishment of new private or municipal knackeries is also prohibited.

According to the competent authorities, the existing processing plant in Luxembourg is fully utilised and sometimes even overloaded.

UNITED KINGDOM

Under the Meat Sterilisation Order, all meat not intended for human consumption must be sterilised. The sterilisation of this raw material (slaughterhouse waste and animal carcasses) is carried out by certain processing firms licensed by the Department of Agriculture.

These processing firms dispose of and collect this slaughterhouse waste. There are no national regulations governing the transport of slaughterhouse waste and carcasses in the UK.

These processing factories (carcass disposal institutions) must pay for their raw material, the price depending on the market position and the costs of production of the meal and fat products.

In general, there are no subsidies for the establishment of carcass processing institutions. However, subsidies are paid regionally and

on different scales for the establishment of carcass processing
institutes and these may be paid up to 20 per cent of the approved
expenditure. The sale of the end products from the carcass processing institutions is direct to fodder manufacturers.

Slaughterhouse waste which cannot be disposed of harmlessly in carcass
disposal undertakings, whether because of inadequate processing capacity or because the distances are too great, must be disposed of by
the authorities. Disposal of carcasses, etc by the authorities is
effected by burying at special places.

There are in the UK still a number of older carcass disposal institutions which do not have a closed system and hence constitute a nuisance to the environment. The UK authorities have in the meantime
closed a number of processing undertakings because of noise and smell
nuisance. The Warren Spring Laboratory has for the past three years
been dealing with the problem of smells from the processing cycle of
a carcass disposal institution.

The Department of the Environment is at the present time preparing
proposals for the authorities as to the best way of solving the
problem.

IRELAND

Despite repeated application and the forwarding of questionnaires,
no details have been received from Ireland.

DENMARK

In Denmark, carcass disposal institutions are partly in the private
sector and partly in the public sector. Under the Carcass Disposal
Order of 4 March 1954, all products supplied to carcass disposal
institutions must be sterilised. This is done in autoclaves, in
which the raw material is sterilised at high pressure and temperature.
The sterilisation temperature is 125^{0}C for a period of 45 minutes.

The carcass disposal institution must be divided up into a clean
department and an unclean department. Dung and waste water from the
institution must also be sterilised in autoclaves before being disposed of. The transport of this waste must be in specially licensed
trucks. After each transport, all parts coming into contact with the
raw material must be disinfected.

The carcass disposal institutions do not receive any public subsidies
worth mentioning. The end products (meal and industrial fat) are sold
on the open market.

The slaughterhouse waste and carcasses are collected and processed by
the 36 carcass disposal institutions existing at the present time.
The burying of carcasses and slaughterhouse waste is not permitted
in Denmark.

9 Prognosis

9.1 FUTURE DEVELOPMENT OF SLAUGHTERHOUSE WASTE IN THE EEC

9.1.1 By type

A change in consumption habits entails a change in the type of slaughterhouse waste. The more affluent the population, the less the demand for processable offal. The maximum economic profit can be obtained if the maximum amount of slaughterhouse by-products goes to human consumption. The function of the slaughterhouse is to seek further possibilities of providing edible by-products for human consumption. The types of slaughterhouse by-products will vary only very slightly.

9.1.2 By quantity

As a result of population growth and increased per capita consumption, the demand for meat and meat products will increase with a parallel increase in livestock, particularly pigs and cattle. Figures 2, 3 and 4 clearly show a consumption rise in livestock. There is a distinct rise in pig stocks in the Federal Republic of Germany, France and Italy, the UK, Denmark, the Netherlands, and Belgium, while the stock of pigs has remained constant in Luxembourg and Ireland. Cattle stock increased in France, the Federal Republic of Germany, the United Kingdom, Ireland, the Netherlands, Denmark and Belgium while in Luxembourg it remained constant and in Italy dropped from 1950 to 1973.

Figures 5 and 6 give the numbers of cattle and pigs slaughtered for the years 1950 to 1973. With the exception of Luxembourg, which has constant numbers of cattle and pigs killed from 1950 to 1973, the figures have increased sharply in the other EEC Member States. Figures 7 and 8 compare the percentage increases in the individual EEC Member States for cattle and pig killings.

The maximum percentage rise in cattle killings (see Survey 7) is in
Italy. The number of cattle killed from 1950 to 1970 rose by 248 per
cent. The other EEC Member States are in the range from 67 to 116
per cent for the years 1950 to 1970. In the years 1950 to 1960 the
percentage rise was in the range from 27 to 79 per cent and from 1960
to 1970 the percentage rise in the EEC Member States was in the range
from 16 per cent (Belgium) to 169 per cent (Italy).

The number of pigs killed (see figure 8) in EEC Member States rose as
a percentage from 11 to 261 per cent from 1950 to 1970. In the period
1950 to 1960 the percentage rise was in the range from 17 to 118 per
cent and from 1960 to 1970 it was in the range from 6 to 146 per cent.

There are still no statistical data for the United Kingdom, Ireland
and Denmark.

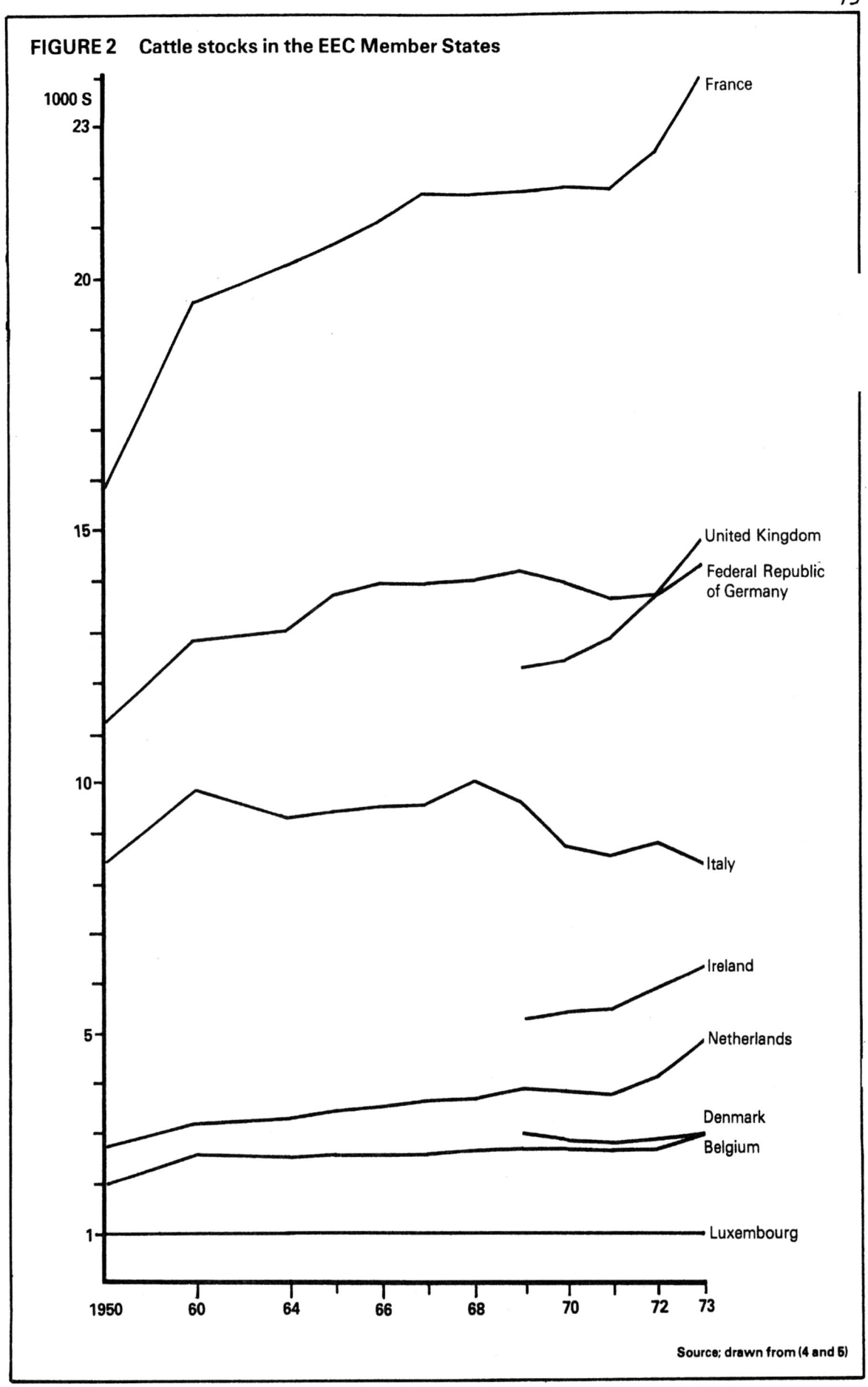

FIGURE 2 Cattle stocks in the EEC Member States

Source; drawn from (4 and 5)

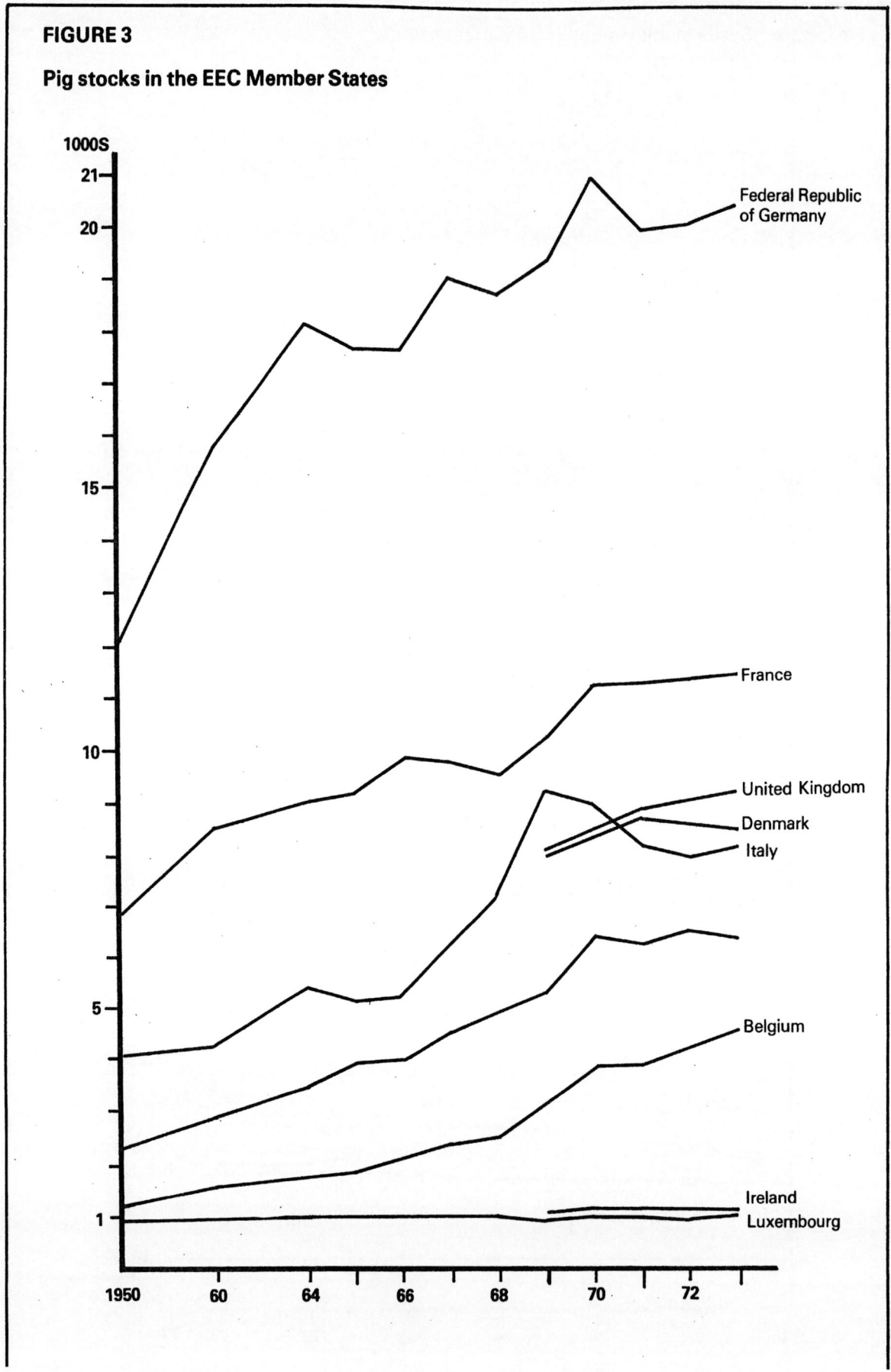

FIGURE 3

Pig stocks in the EEC Member States

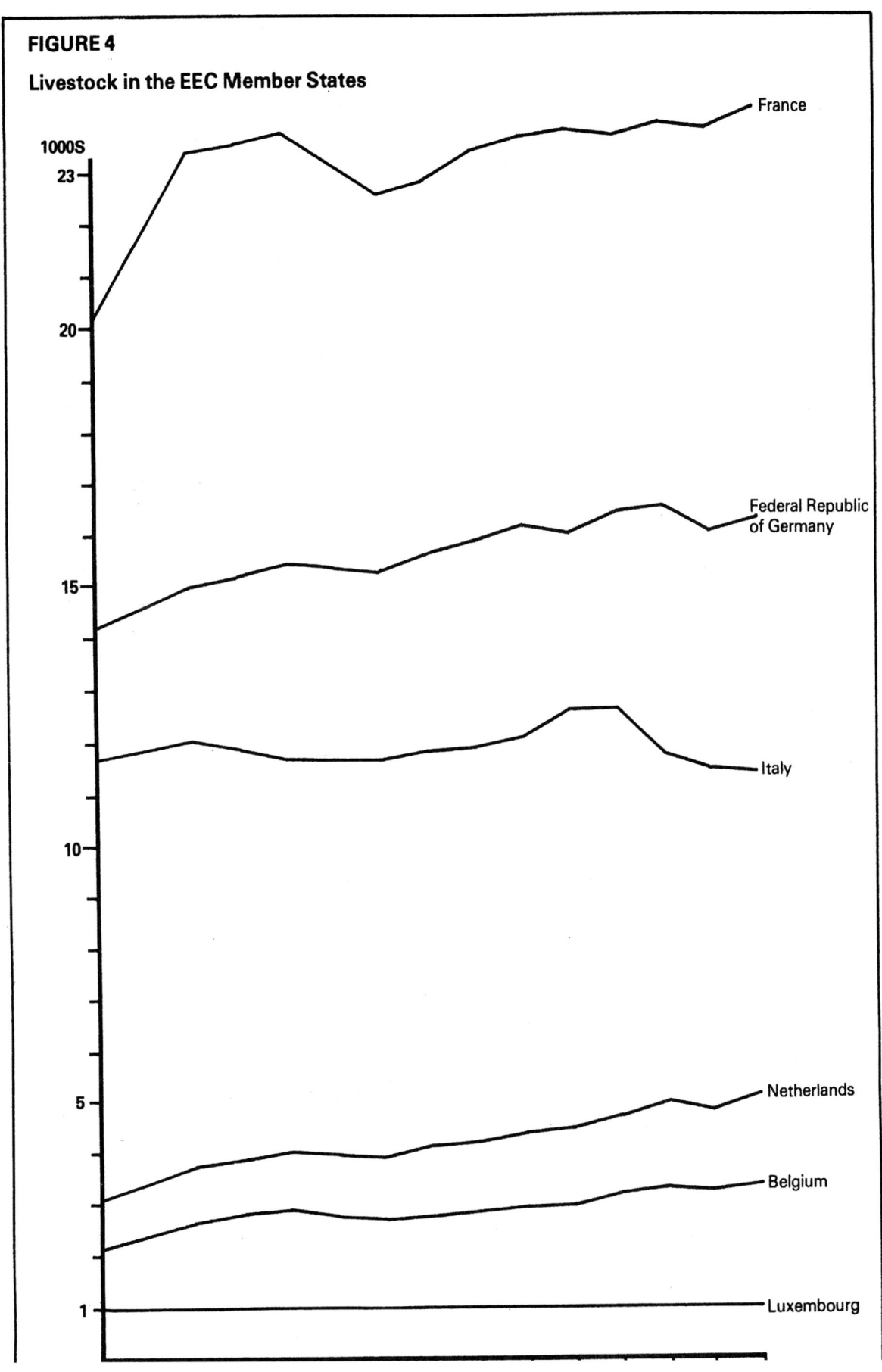

FIGURE 4

Livestock in the EEC Member States

FIGURE 5

Number of animals slaughtered in the EEC Member States

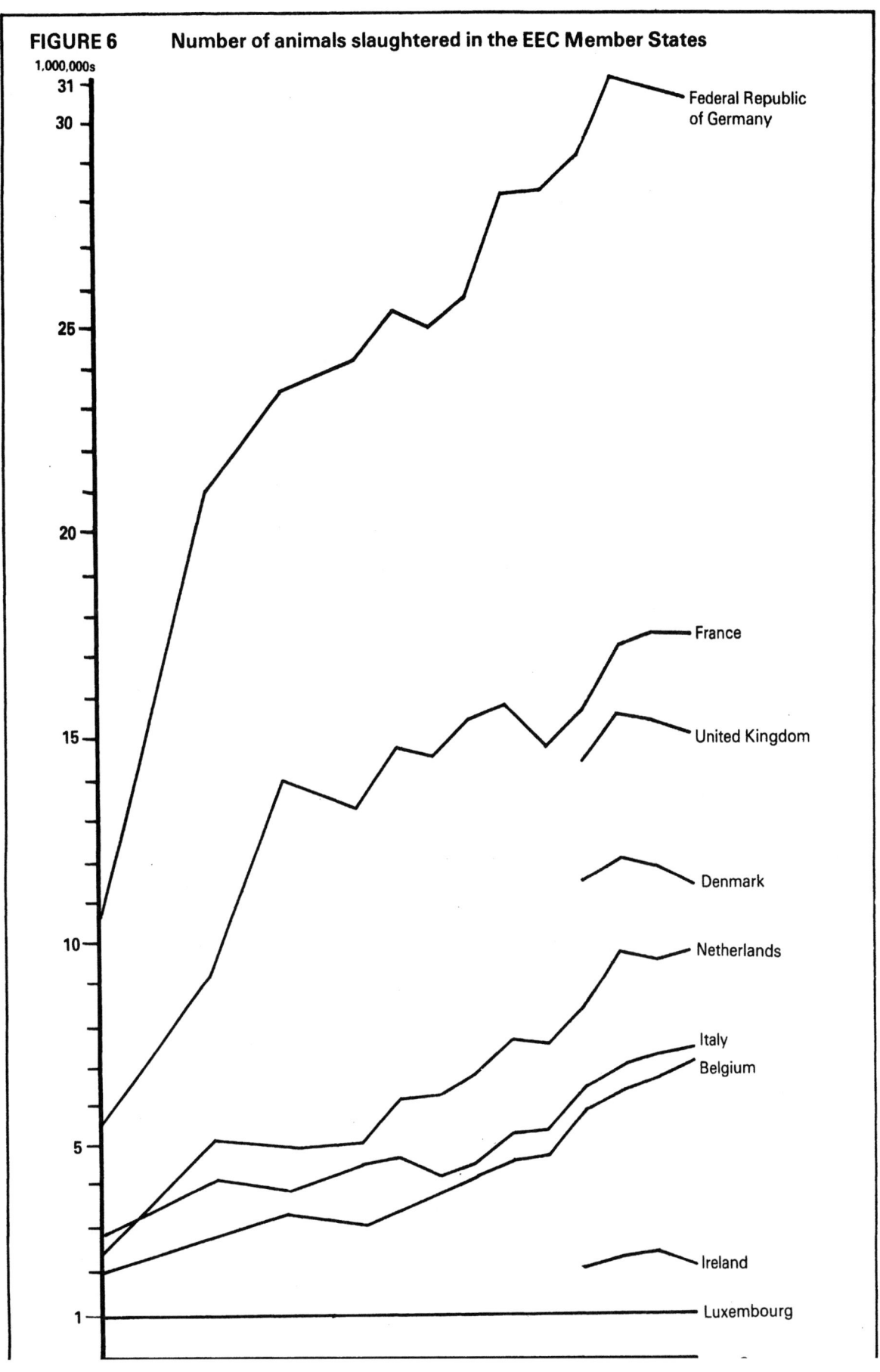

FIGURE 6 — Number of animals slaughtered in the EEC Member States

FIGURE 7

Increase in the number of animals slaughtered in the EEC Member States from 1950 to 1970

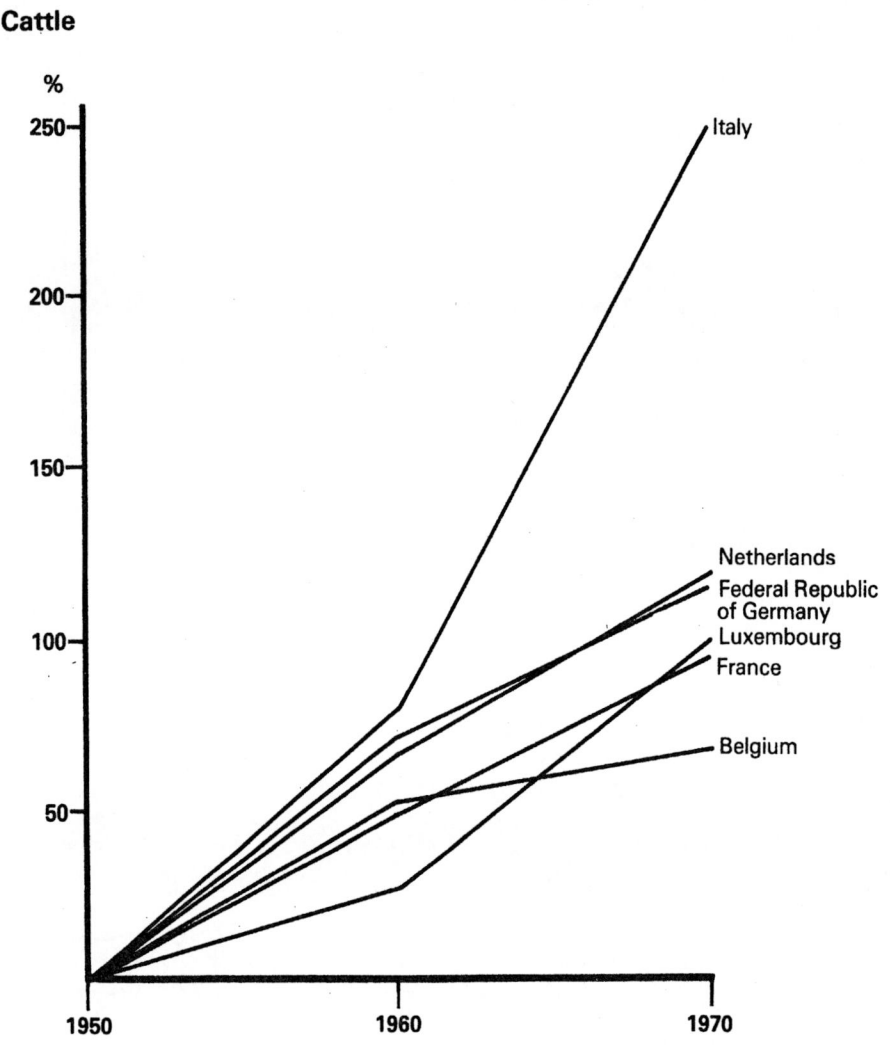

Cattle

Table 28 (in '000s 1950 = 100)

Federal Republic of Germany	2.142	3.661 = 170%	4.628 = 216%
France	2.019	3.025 = 149%	3.917 = 194%
Italy	1.120	2.016 = 179%	3.903 = 348%
Netherlands	422	702 = 166%	925 = 219%
Belgium	458	692 = 151%	766 = 167%
Luxembourg	22	28 = 127%	44 = 200%
United Kingdom			3.683
Ireland			899
Denmark			488

FIGURE 8

Increase in the number of animals slaughtered in the EEC Member States from 1950 to 1970

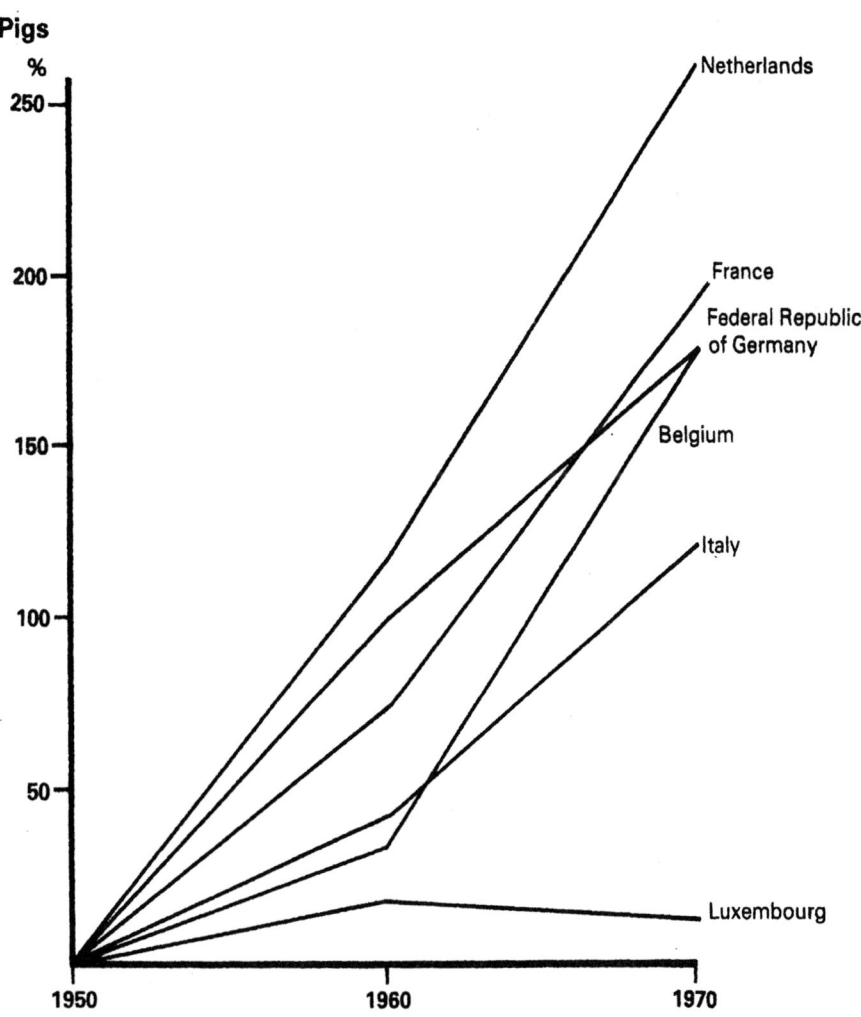

Table 29 (in '000s 1950 = 100)

Federal Republic of Germany	10.464	20.951 = 200%	29.220 = 279%
France	5.325	9.283 = 174%	15.703 = 294%
Italy	2.878	4.102 = 142%	6.383 = 221%
Netherlands	2.344	5.116 = 218%	8.468 = 361%
Belgium	2.066	2.750 = 133%	5.782 = 279%
Luxembourg	114	134 = 117%	127 = 111%
United Kingdom			14.391
Ireland			2.102
			11.497

Table 30 shows the average per capita meat consumption in kg per annum. A comparison of these figures will show that France has the highest beef and veal consumption and also the highest poultry consumption per capita of the population.

The Federal Republic of Germany has the highest pork consumption.

In respect of total meat consumption, France is at the top with 96 kg in 1970, followed by the Federal Republic of Germany with 87 kg, Ireland with 84 kg, Belgium and Luxembourg with 83 kg, the UK with 72 kg, the Netherlands with 66 kg, Denmark with 63 kg and Italy with 57 kg.

These figures clearly show that most EEC countries have still not reached saturation point.

The total meat consumption is also dependent on the population growth (see table 31) and on the ever-increasing per capita income.

All these factors increase meat consumption and hence slaughterhouse waste increases to the same degree.

Under the given conditions it may be assumed that the increase in slaughterhouse waste in the EEC area for a ten year period will be in the range from 30 to 50 per cent, the increase being due less to an increase in per capita consumption than to population growth in the countries which already have a per capita consumption of more than 80 kg of meat (Federal Republic of Germany, France, Belgium, Luxembourg, Ireland). In the case of those countries (Italy, Netherlands, UK, Denmark) which have much less than an 80 kg per capita meat consumption, the increase in meat consumption will be due firstly to the population growth and secondly to the increase in per capita consumption, and 100 kg may be regarded as the top limit. The amount of slaughterhouse waste will increase to the same degree as the increase in meat consumption.

81

Table 30: Average per capita meat consumption in the EEC area

Units : kg/year

Product	Year	FRG	France	Italy	Netherlands	Belgium/Luxembourg	UK	Ireland	Denmark
Beef and veal	1968/69	24	31	23	22	26			
	1969/70	24	30	24	21	27			
	1970/71	25	30	25	22	28	21	19	20
Pork	1968/69	44	30	11	29	31			
	1969/70	44	30	11	27	32			
	1970/71	48	31	13	30	34	23	31	30
Poultry meat	1968/69	7.4	13.7	10.4	5.2	7.7			
	1969/70	7.9	14.6	11.2	5.2	7.9			
	1970/71	8.6	14.9	11.6	6.0	8.7	10.4	9.9	5.0
Total meat	1968/69	80	94	50	62	76			
	1969/70	81	94	54	61	80			
	1970/71	87	96	57	66	83	72	84	63

Source: (6)

Table 31: Population of the EEC

Units : '000

Year	FRG	France	Italy	Nether-lands	Belgium/Luxembourg	UK	Ireland	Denmark
1954/55	52,127	43,228	48,768	10,680	9,129			
1955/56	52,698	43,627	49,191	10,822	9,180			
1956/57	53,319	44,059	49,052	10,957	9,240			
1957/58	53,994	44,563	49,311	11,096	9,302			
1958/59	54,606	45,015	49,640	11,278	9,359			
1959/60	55,123	45,465	50,023	11,417	9,409			
1960/61	55,785	45,904	50,372	11,556	9,458			
1961/62	56,473	46,422	50,672	11,721	9,511			
1962/63	57,029	47,573	50,996	11,890	9,574			
1963/64	57,581	48,059	51,358	12,042	9,651			
1964/65	58,195	48,562	51,778	12,212	9,738			
1965/66	58,792	48,954	52,109	12,377	9,811			
1966/67	59,174	49,374	52,443	12,535	9,868			
1967/68	59,310	49,728	52,457	12,661	9,909			
1968/69	59,748	50,107	53,061	12,798	9,939			
1969/70	60,352	50,522	53,399	12,958	9,965			
1970/71	61,001	51,012	53,722	13,119	9,996			
1971/72	61,503	51,487	54,067	13,260	10,025	55,678	2,994	4,976

Source: (4)

9.2 FUTURE DEVELOPMENT OF LEGISLATION AND ADMINISTRATIVE REGULATIONS

It is apparent from the past that there has been considerable risk to both human and animal health from dead animals and slaughterhouse waste which was not disposed of or processed properly. The risk of the spread of bacterial and parasitic pathogens continually increases because of the decreasing space available to man and animal with the simultaneous rise in waste from animal production. The reason for the increasing risk of the spread of such pathogens is that toxic degradation products form on the degradation of organic material. The legislators have recognised these risks and therefore drafted appropriate laws and regulations.

In practically every EEC country laws and regulations are under discussion to allow for the above problem. The main focus of attention has been the hygienically satisfactory disposal of slaughterhouse waste and economic processing for further use.

10 Alternative solutions

10.1 POSSIBILITIES OF PROCESSING SLAUGHTERHOUSE WASTE BY TYPE AND QUANTITY

All slaughterhouse waste can be processed industrially in some way or other. The following chapters show the possibilities of processing slaughterhouse waste for industrial and agricultural use.

10.1.1 In industry

Types:

Hides and skins

The term 'hide' is used to denote the integument (outer skin) together with the hair covering of fully grown cattle and horses.

The term 'skin' is used to denote the integument of young slaughtered animals and sheep. These types of waste are the most valuable slaughterhouse waste and are without exception processed industrially. The following are industrial users:

    Leather goods industry
    Furriers
    Gelatine industry
    Glue industry

Some 65 to 70 per cent of the hides and skins are processed into leather. The remainder are processed as offcuts, partly in the gelatine industry and partly in the glue industry. This waste, however, is not obtained until after tanning. Small quantities also pass to the carcass disposal institutions, where they are ground into animal meal.

Animal hair and bristles

The animal hair generally remains on the hides and skins. Skin hair occurs as a by-product from tanning works. The brushes and paintbrush industry is one possibility for industrial processing. Pig's bristles can be used in the brushes, paintbrush and upholstery industry.

Claws and horns

Claws and horns are today practically unused industrially. Combs and buttons were previously made from claws and horns. However, they are partly used as an additive to the extinction agent in fire extinguishers, for which purpose they are dried and ground. Horn shavings are used as organic fertilisers.

Lungs, spleen, udder, stomach (including rumen), trachea and chitterlings

If these parts are not condemned they can be used in the petfood industry. In most cases, however, they pass to the carcass disposal institutions which make carcass meal from them.

Contents of stomach and intestines

There is practically no industrial processing for this type of by-product. Disposal should be as manure together with the faeces and litter and care should be taken to prevent this waste from entering the sewers.

Blood

Blood is of only minor importance in human nutrition. Some 10 to 15 per cent of pig's blood is processed into sausage products. A further possible process would be to obtain plasma from blood. This blood plasma may be used in meat and frankfurter type sausage production. Blood meal can also be made from blood by drying and be used for animal feed. It is also possible to prepare wood glue from blood by certain technical processes.

Intestines

The intestines have long been processed into sausage casings but this type of utilisation of the intestines is of interest only in those countries where there is a very low wage level since this type of processing is very wage-intensive.

Most of the intestines are processed into meat meal at the animal carcass disposal institutions.

Bones

It is a difficult matter to determine quantities of bones. The bones occurring at slaughterhouses, butchers and similar undertakings may be collected by the bone-processing undertakings. A considerable proportion of the bones goes to the final consumer either with the meat or in the form of soup bones. These bones are lost to any further processing. The bones are used in the glue and gelatine industry. Only bones which have been boiled down are suitable for glue production. Bones that have been in an autoclave are not suitable. The gelatine obtained from the bones is used in the photographic industry, the typewriter stencil industry and in printing works.

The following can be made from bones for animal feed purposes:

1. Meat and bone meal
2. Coarse feeding bone meal
3. Feeding bone meal
4. Bone ash
5. Bone charcoal
6. Calcium phosphate
7. Bone fat

Bone charcoal is not only used for processing as a fodder but also in the sugar industry for bleaching sugar, in the air filter industry, and in the pharmaceutical industry, in the chemical industry as a dyestuff for dyeing, in the shoe polish industry for making finishes, and last but not least as a filler for rubber products.

Bone fat, meat and bone meal, feeding bone meal, calcium phosphate, and coarse feeding bone meal are added to mixed fodders, giving a guaranteed 24 per cent raw protein and 8 per cent phosphorus approximately.

Slaughterhouse fats

Slaughterhouse fats not used for human consumption are processed in fat and tallow works and by pork fat processors, and supplied to the appropriate industries.

Industrial fats

These occur in the TBA processes and are used in the chemical industry mainly for the manufacture of detergents.

Glands

Glands are processed by the pharmaceutical industry which requires the secretions from the glands for the production of drugs and sera. The drug companies frequently arrange their own collection transport.

All the other slaughterhouse waste is processed into carcass meal in the carcass disposal institutions.

Processing of slaughterhouse waste by quantity

The processing of slaughterhouse waste in the relevant industries as listed above is economic only if there are sufficient quantities available and a continuous supply is ensured. Transport costs also play a considerable part in processing. In slaughterhouses which do not have an adequate supply and perhaps an unfavourable position, no differentiation is made between the individual types of slaughterhouse waste. In such undertakings, mainly the medium-sized and small ones, all the waste is assembled in a condemned meat area and is collected by the carcass disposal institutions at specific intervals. A considerable amount of slaughterhouse waste is also disposed of via the sewers at these undertakings. This applies, for example, to blood, meat and bones, the stomach, and intestine contents, etc (see figure I).

10.1.2 In agriculture

In general, all slaughterhouse waste can be returned to agriculture via the carcass disposal institutions, which make carcass meal from this waste. With the exception of the stomach and intestine contents and the faeces together with litter, which is used as an organic fertiliser, it is not possible to utilise slaughterhouse waste directly in agriculture because of the risks of infestion.

The skin, bristles, claws and horns, which are processed by the carcass disposal institutions, must be screened out of the carcass meal. The screenings are used as organic fertiliser because of their high keratin content (eg horn meal or horn shavings).

Poultry waste can be utilised direct in agriculture if certain measures are taken.

The waste (gullet, trachea, lungs, intestines, sex organs, stomach and intestine contents, head, shanks and webbed feet, which make up 14 per cent of the live weight) can be fed direct to fattening pigs after boiling and chopping up. The blood can also be fed directly in the form of cooked blood or silage. On the other hand, the feathers have to be processed into feather meal in a special processing plant to be usable as fodder.

Generally all blood can be used for feeding in agriculture provided it is by way of silage.

10.2 DISPOSAL OF SPECIAL WASTE

Those carcasses and carcass parts which are unfit for human consumption because of disease and toxicosis are termed special waste. A distinction must be made between diseased animals and poisoned animals in connexion with processing.

Diseased animals can be processed by the carcass disposal institutions, but poisoned animals must not be so processed, and must go to an

incineration plant. Another possibility is to dispose of poisoned
carcasses and carcass parts harmlessly in a digestion tower.

10.3 PROCESSES FOR MAKING SLAUGHTERHOUSE WASTE RE-USABLE

As will be seen from table 4, the processing of slaughterhouse waste
comes under three broad headings:

1. Waste water
2. Special processing
3. Carcass disposal institutions

The significance of the individual disposal systems will now be
described briefly.

1. Waste water

Waste water is generally liquid water, which may contain large quantities of pathogens. This waste water may cause endemic or epidemic
diseases in man and beast. Salmonella may cause paratyphus.

A number of types of salmonella occur in waste water. The contamination of waste water with salmonella is due primarily to slaughterhouses, butchers, fish and sausage factories, the meat-processing
industry, carcass disposal institutions, sausage casing processors,
gut string manufacturers, dairies and cheese-makers.

Slaughterhouses and cattle farms produce considerable quantities of
waste water depending upon:

1. the type of livestock for slaughter;
2. the type of operational equipment;
3. the volume of waste washed away;
4. the total weight of the slaughtered livestock.

Strauch[15] puts the average waste water at 2.26 to 10.1 m³/t live
weight. All the waste water from a slaughterhouse and cattle farm is
infectious and must be regarded as a source of risk.

Such waste contains blood, bristles, urine, faeces, bone splinters,
meat residues, fats, and the contents of the stomach and intestines
of the slaughtered animals.

Waste water from carcass disposal institutions contains the vapour
condensate of the boiler, the rinsing, washing, flushing and percolating water, cooling water and rain water.

This waste water is unobjectionable from the disease aspect, if it is
carefully disinfected and processed.

If it is not sterilised, anthrax spores may occur in the waste water
from carcass disposal institutions and endanger man and animals for
many years.

10.3.1 *Processes for making certain waste re-usable*

1. In waste water

Mechanical processes

a) Oil and fat trap plants

Before the waste water is discharged to the communal sewer system,
the oil and fats should be removed in order to avoid loading the
communal clarification plant and interfering with its operation. The
fat is removed from slaughterhouse waste water by reducing the rate
of flow, oils and fats floating on the surface in such conditions.

A second possibility would be to aerate the waste water from the
bottom of the tank, small air bubbles floating up and entraining the
fats from the water. In these conditions the fats form a foam which
passes over weirs into special scoop chambers.

Recently, separators have frequently been used for screening solids from slaughterhouse waste water[20].

b) Mechanical cleaning

On the sedimentation principle, all substances capable of sedimentation settle within a period of two hours. This method enables 30 per cent of the total foreign substances to be separated from the water. In the sedimentation process, most of the eggs of the bacteria and parasites settle into the sludge so that the latter has a very high pathogen concentration.

The separated sludge, which contains about 95 per cent water, is objectionable from the disease aspect and must therefore be stabilised, ie the sludge must be brought into a state such that it does not give rise to any environmental pollution. The processes used for this purpose at present are as follows:

1. Anaerobic treatment (ie with the exclusion of air) ▪ sludge digestion
2. Heat treatment.

The anaerobic digestion process is carried out in closed unheated or heated digestion chambers or towers. The micro-organisms give rise to a biological process in which gas is liberated. This gas has a calorific value of 6,000 to 8,000 heat units with the following composition:

75 to 90% methane
10 to 35% carbon dioxide
0 to 5% water
0 to 10% nitrogen
0 to 0.5% hydrogen sulphide

The gas can be used within the undertaking. Sludge which has been subjected to an anaerobic treatment (digestion process) is not unobjectionable, however, from the disease aspect. The water content,

95 per cent, is also very high. Another step would be to dewater the digested sludge. To obtain a product which is satisfactory from the disease aspect, the air-dried clarified sludge can be composted.

The heat occurring during composting ensures that the clarified sludge is hygienised. A favourable effect is obtained from the fact that this dewatered clarified sludge can be composted in conjunction with dung and domestic refuse.

The compost produced in the composting factory can be used in agriculture, nurseries, for landscaping, and also for the industry for making slabs and other products (see figure 9).

A new process enables the clarified sludge to be directly composted[16] with temperatures occurring in the region of 80°C, which results in disinfection of the material. This type of direct clarified sludge composting is economic if there are large quantities of clarified sludge which cannot be processed together with domestic refuse. This may be because of unequal conditions or because the position of the clarified sludge can be co-ordinated with that of the composting works. In the case of a new installation for a composting plant, care should be taken to ensure that the ratio of clarified sludge to domestic refuse corresponds to the population equivalents. Clarified sludge and domestic refuse as a compost base can be relatively satisfactorily composted because the C:N ratio can be compensated.

Both fresh sludge and digested sludge are suitable for heat treatment. However, the sludge does not have to be fully digested for heat treatment. The term 'heat treatment' is used to denote incineration of the sludge. Fresh sludge has a higher calorific value than digested sludge and is more suitable for incineration. With this method, all the pathogens in the fresh sludge are killed. However, digestion of the sludge followed by composting (see 12.3 for a fuller discussion) is much cheaper and is more conservatory because heat treatment requires considerable quantities of energy.

94

The fats obtained from the fat trap plant can be supplied to the
fat-processing industries.

Biological processes

Not all foreign and noxious substances can be removed from the waste
water by mechanical cleaning. The second purification stage is the
biological waste water purification process, in which purification
takes place by biological degradation processes. The biological
purification takes place aerobically. A distinction is made between
the natural process and the synthetic process. The natural processes
include sewage farms, sprinkling plants and fishpond processes.

The synthetic processes include the percolating filter and activated
sludge processes.

The second purification stage (biological) is not economically viable
for waste water purification at slaughterhouses. The biological process is used mainly for communal sewage purification.

The mechanical processes are intended to provide pre-purification of
the slaughterhouse waste water before it is discharged into the communal sewer system where it passes through the second purification
stage.

2. *Special processing*

As already explained in detail in previous chapters, the term special
processing is used to denote the processing of slaughterhouse waste
which is used in industrial undertakings apart from carcass disposal
institutions. This slaughterhouse waste includes hides and skins,
animal hair and bristles, claws and horns, lungs, spleen, udder,
stomach, rumen, blood, intestines, bones, slaughterhouse fats and
glands. The processing of these products in special undertakings
depends greatly on the amount of waste accumulating and transport
costs. Without exception, processing in these industries is carried

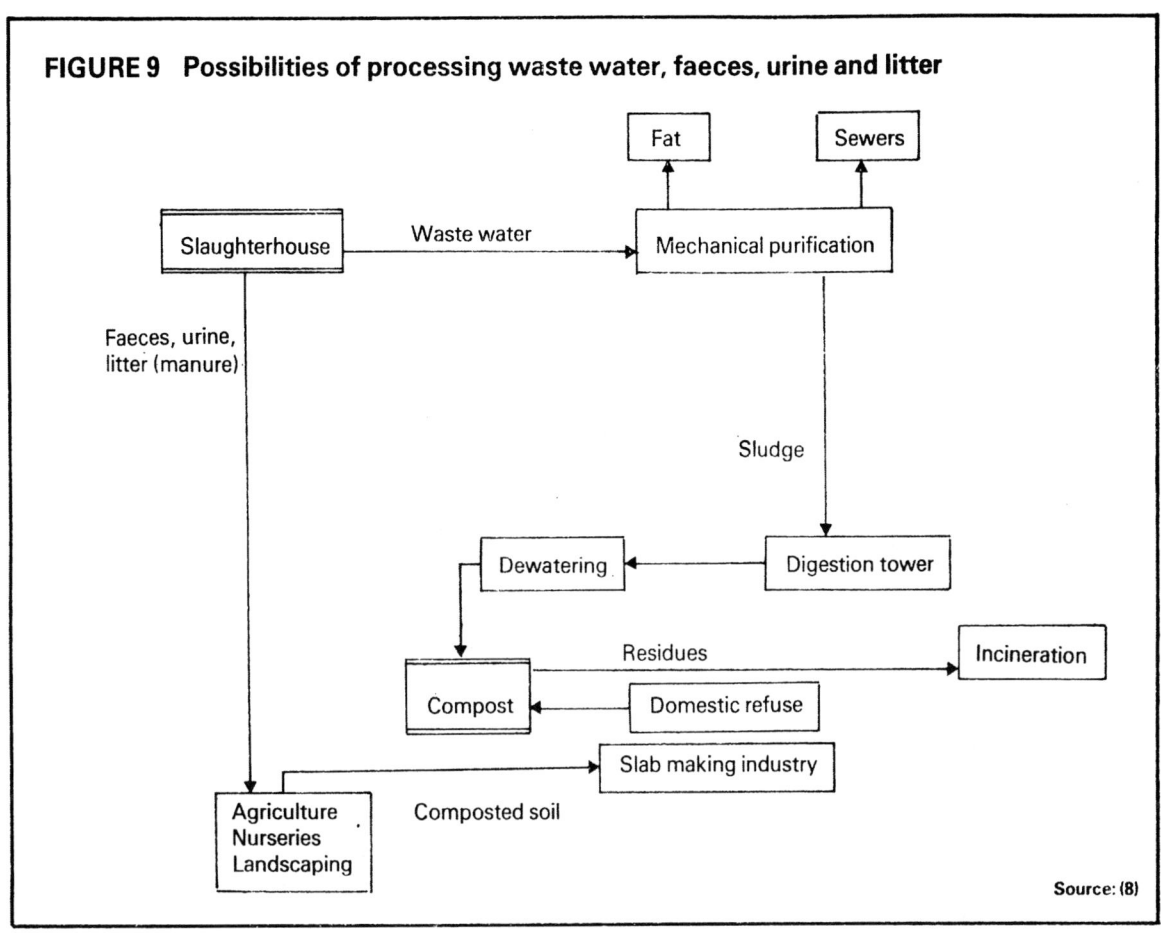

FIGURE 9   Possibilities of processing waste water, faeces, urine and litter

Source: (8)

out only from economic aspects. If it is not economic to process
these special products, then the further processing is carried out
by carcass disposal institutions. For this reason, the individual
processes will not be discussed in detail in this study.

3. *Carcass disposal institutions*

These are an important link in the chain for maintaining public
health and general hygiene. They are responsible for the collection
and harmless disposal of dead animals, condemned meat and waste from
slaughterhouses, dairies and the fish-processing industry. In carrying out their task they must not only bear in mind economic considerations, but must also give priority to hygienic measures (see
figure 10).

The processes used are based fundamentally on harmless disposal of
the waste under pressure and high temperatures for a certain period,
the end products being meal on the one hand and industrial fats on
the other.

96

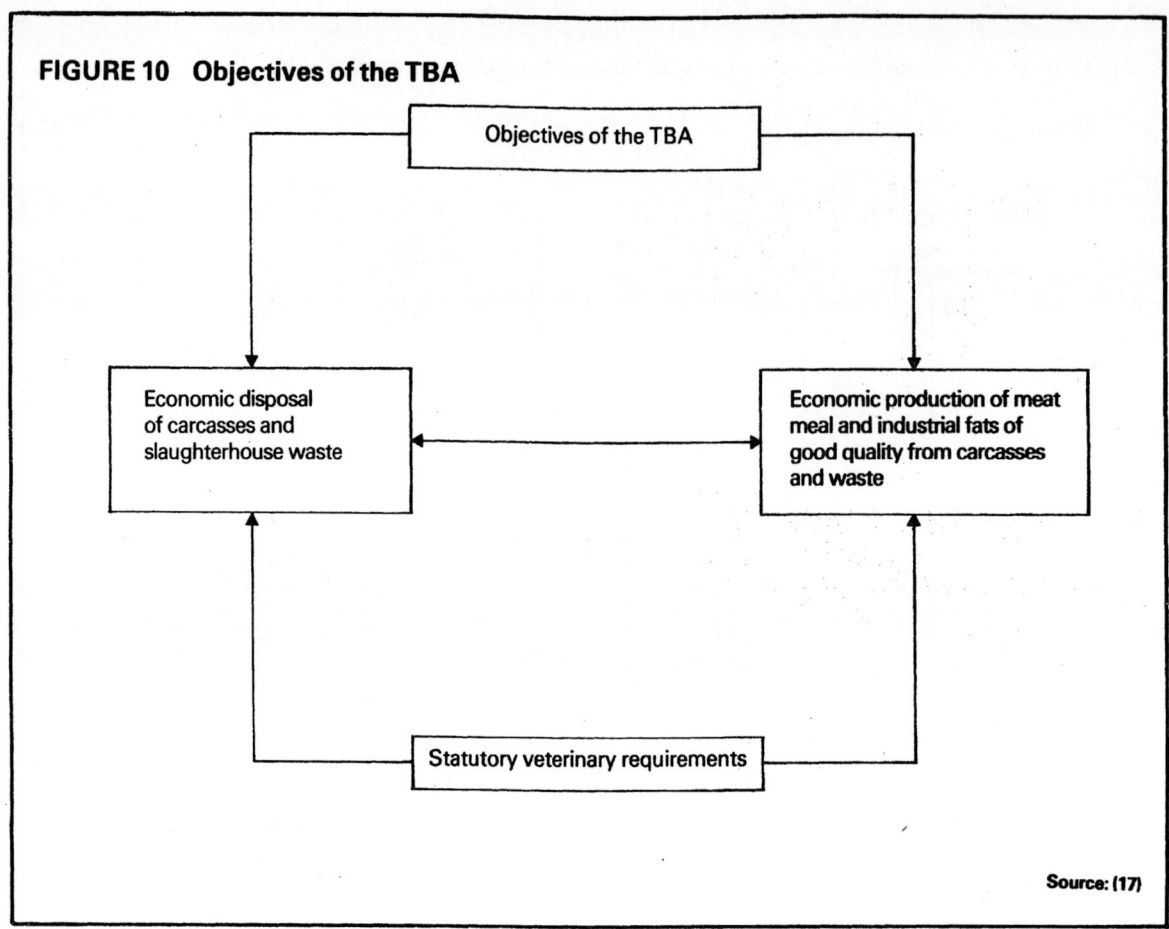

FIGURE 10   Objectives of the TBA

Objectives of the TBA

Economic disposal
of carcasses and
slaughterhouse waste

Economic production of meat
meal and industrial fats of
good quality from carcasses
and waste

Statutory veterinary requirements

Source: (17)

The most conventional processes are as follows:

1. Dry process
2. Perchloroethylene process (or azeotropic process or per process)
3. Wet process.

The wet process is very little used today because the end products are of lower quality. In contrast to the dry process, the wet process operates directly with steam for sterilisation and disintegration. The products are usually animal meal, animal fat and glue liquor.

Since this method is only very restricted in use, it may be disregarded here.

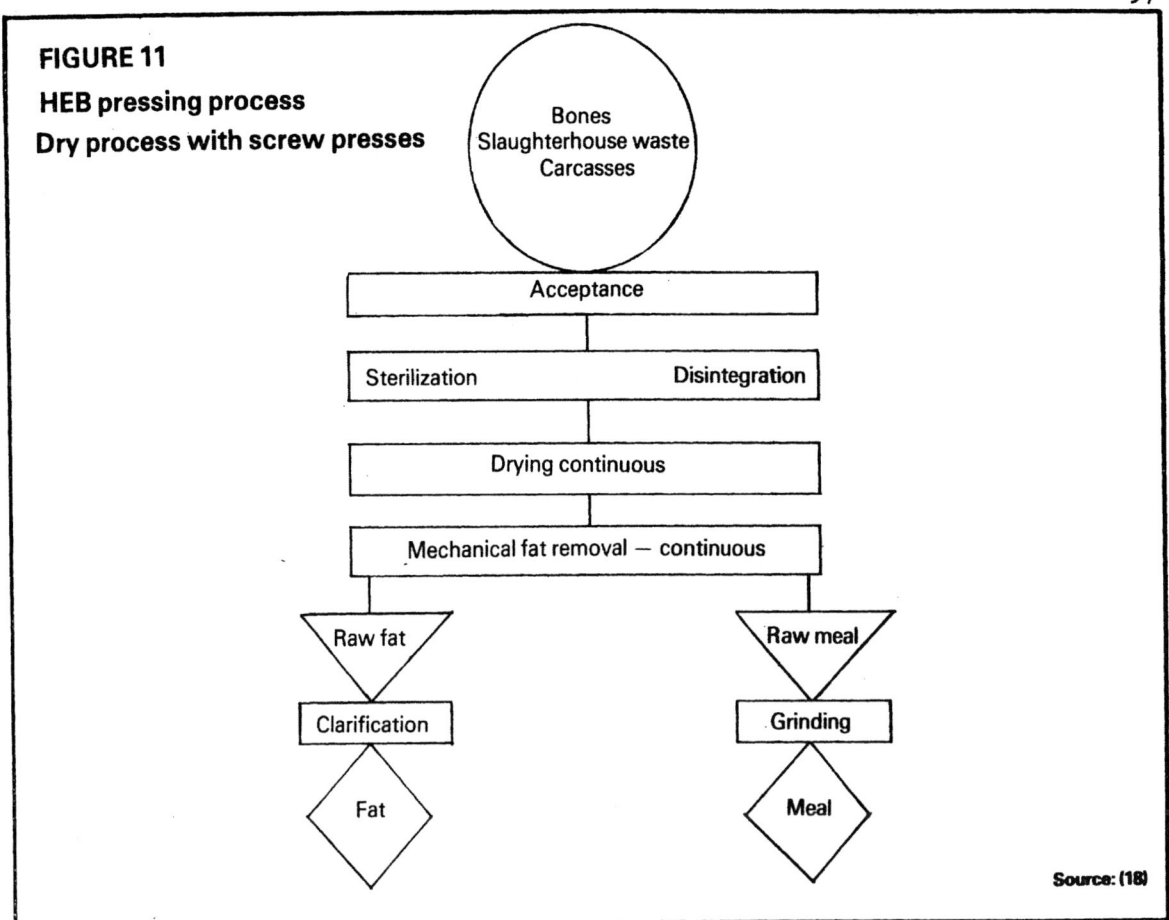

FIGURE 11
HEB pressing process
Dry process with screw presses

Bones
Slaughterhouse waste
Carcasses

Acceptance

Sterilization   Disintegration

Drying continuous

Mechanical fat removal — continuous

Raw fat   Raw meal

Clarification   Grinding

Fat   Meal

Source: (18)

The dry process

This process is the most widespread. It is also know as the pressing
process. The fat is removed from the meat pulp by means of screw
presses. The residual fat content of the meal produced is between
7 and 11 per cent. Depending upon the size, the output of this process is between 60 and 250 y/24-hours (see figures 11 and 12).

The 'per' process

The fat-containing material is removed by perchloroethylene or benzine.
Benzine was previously used for disintegration instead of perchloroethylene. The process is hardly ever used with benzine today because
of the high risk of explosion. Perchloroethylene is used instead and
there is a distinction between two processes:

a. the wet extraction process (see figures 13 and 14)
b. dry extraction (see figures 15 and 16).

The difference between the two processes is that the fats are extracted
immediately after sterilisation in the case of wet extraction while

In the dry extraction process the sterilised homogeneous flowable
meat pulp is dried indirectly. After this drying operation the fat
is removed by a solvent (perchloroethylene). With the wet extraction
process the residual fat content is 4 per cent while with the dry
extraction process the residual fat content is generally between 3
and 7 per cent. (See description of the individual processes).

The fats produced, which are known as industrial fats, are used
mainly in the chemical industry for the manufacture of detergents.
The meal is used in agriculture for animal feeds.

Description of the HEB pressing process (see fig 12)

With this process, large processing capacities can be installed in
the minimum floor space. Only one man is required for supervision,
irrespective of the plant capacity. The continuous operation of the
dryers avoids uneconomic steam peaks. The HEB pressing process is
subdivided into three main grpups:

A.   Charging of raw material, sterilisation and bunkering
     of the material
B.   Continuous drying
C.   Continuous fat removal

A. Charging of raw material, sterilisation and intermediate bunkering of the material

The boiler plant shown diagrammatically in the drawing is intended
to receive and sterilise the carcasses and slaughterhouse waste.
The raw material bunker with the feed screw (1) is directly connected
to the boiler and steriliser (2). This ensures hygienic reception
of the raw material in the process.

The boiler has a large filling aperture which enables even carcasses
which have not been cut up to be introduced without any problems.
The equipment is normally sealed by a special valve operated by an
electric motor.

99

FIGURE 12

HEB pressing process

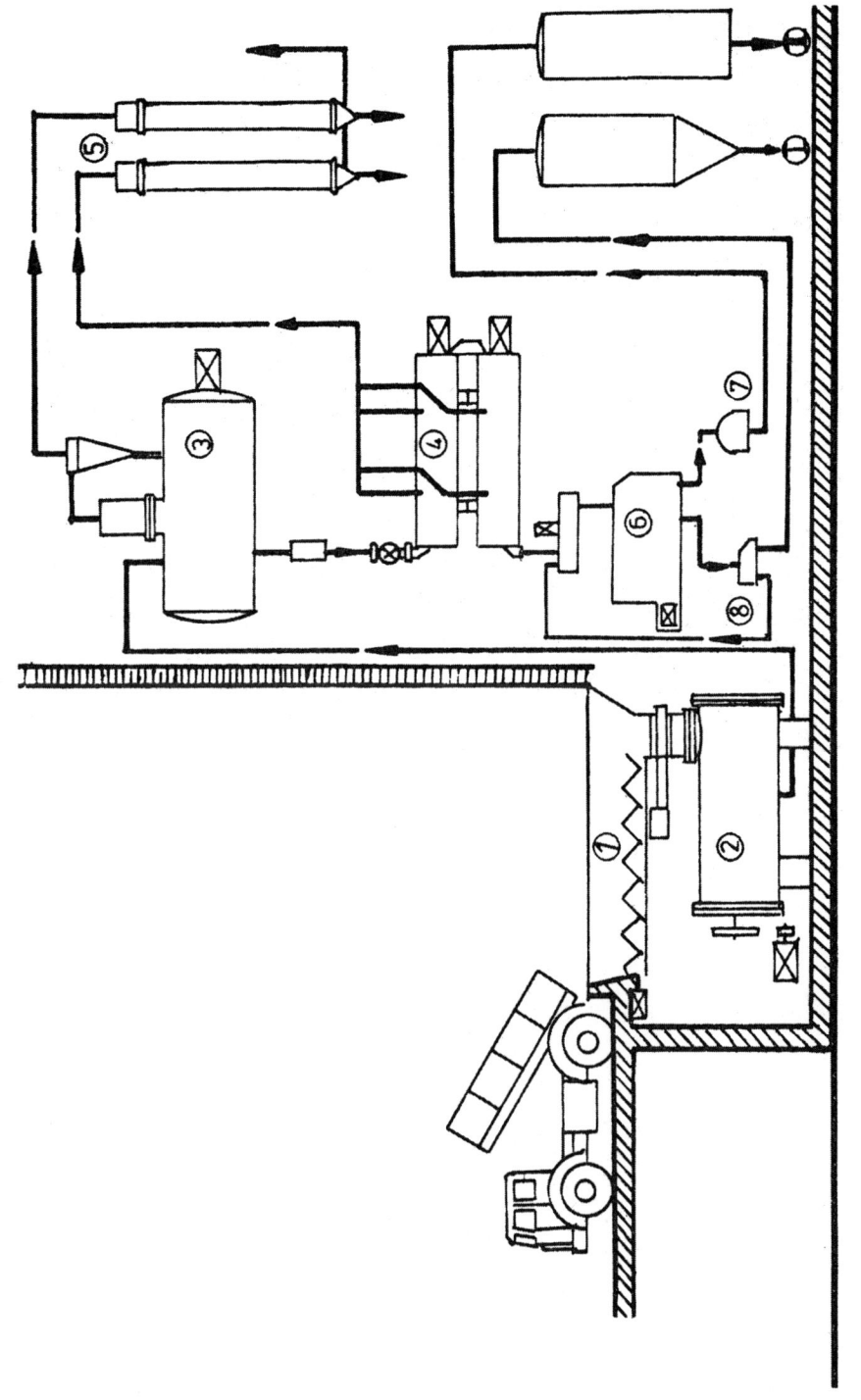

1 Bunker
2 Boiler and sterilizer
3 Intermediate bunker
4 Drying plant
5 Condensers
6 Screw press
7 Meal processing
8 Fat processing
I Fat
II Meal

Pressing process

Source: (18)

The special agitator construction of the boiler enables the raw material to be comminuted partly mechanically and partly thermally, without foreign bodies causing any damage.

On completion of sterilisation, the material is forced into the intermediate bunker (3) by means of the internal pressure.

The advantages of this boiling and sterilisation plant are that the raw material is emptied from special vehicles directly into a large-volume chute which can be covered.

Uniform mechanical and thermal disintegration of the material to the required particle size.

All the foreign bodies, such as iron and pieces of wood and other parts which cannot be disintegrated, remain in the boiler without causing any damage.

B. Continuous drying

The dimensions of the drying plant (4) and associated condensers (5) are such that they can be used in practically any undertaking. The output is variable. The design takes into account experience gained in respect of corrosion and mechanical abrasion, by using corrosion-resistant and wear-resistant materials.

The process is controlled firstly by controlling the entry of material to the dryer and secondly by controlling the flow of material from stage to stage. The plant can be operated with steam to a pressure of eight atmospheres gauge. The entire process from the boiler on takes place continuously and indicators, control and alarm equipment are provided.

C. Continuous fat removal

The screw presses (6) are adapted to the capacity of the preceding drying plant and contribute to the continuous removal of fat from the dried meat. Depending on the composition of the raw material and its state of freshness, the residual content of the meal produced is in the range from 7 to 11 per cent.

Comminuting machines (7) and special fat processing devices (8) are also included in the overall plant described here.

Description of HEB wet extraction

A. Applications

The wet extraction process is suitable for processing all organic waste substances from slaughterhouses and is used in the following:

1. Carcass disposal institutions
   Raw material: carcasses, offal, bones in any composition and small proportions of blood and waste hide

2. Bone extraction installations
   Raw material: bones, raw fats

3. Slaughterhouses
   Raw material: all the raw materials as listed under 1 and 2 in varying composition

4. Poultry slaughterers
   Raw material: waste, feathers, blood - by separate processes.

General details:

With the wet extraction process it is possible to process slaughterhouse waste of all kinds, condemned meat, bones, blood and carcasses, mixed or separately, in one operation, to recover all the solids contained in the raw material and to form high-grade end products, allowing for the following requirements:

102

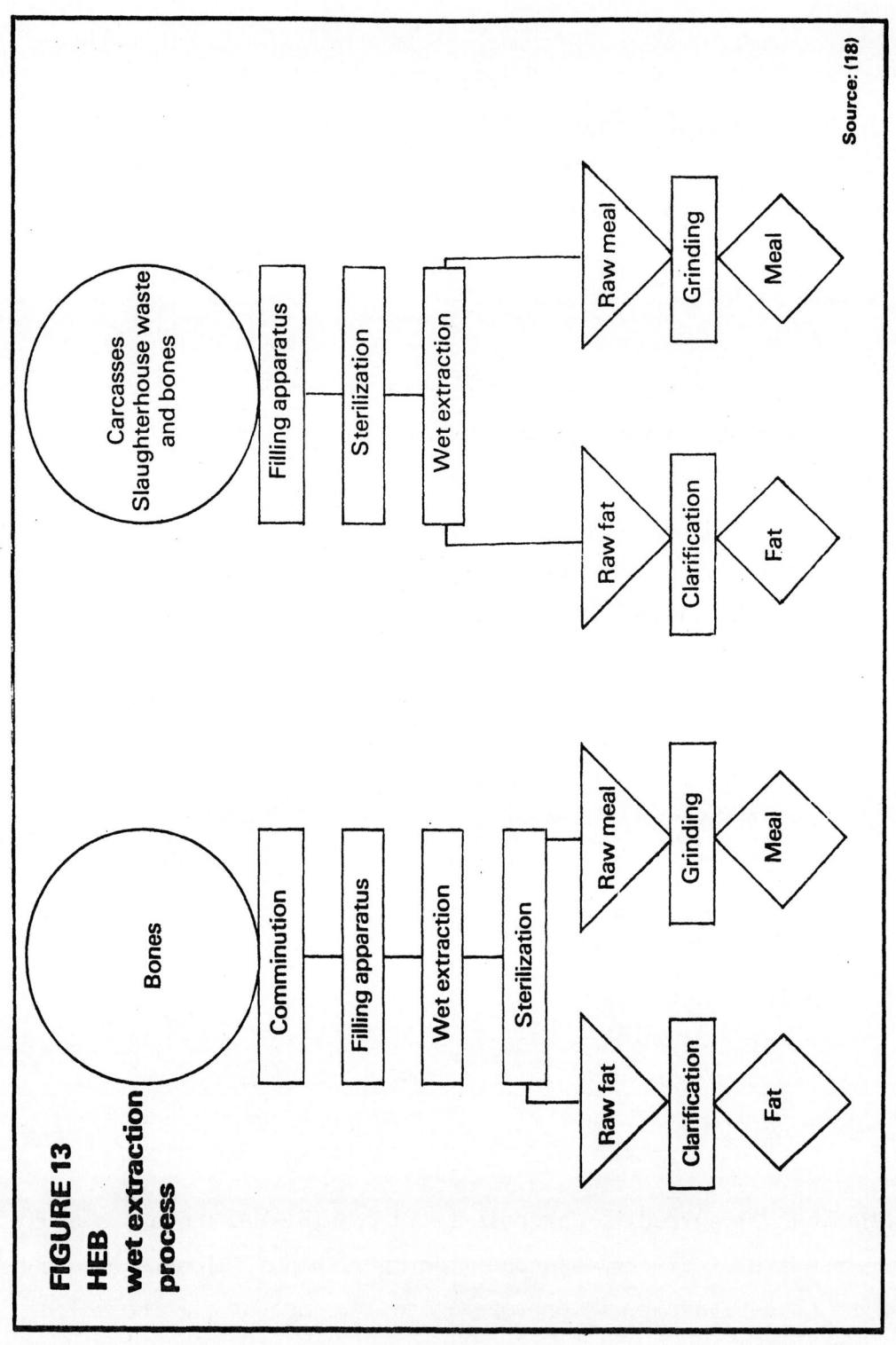

FIGURE 13
HEB
wet extraction process

Source: (18)

a)   completely odourless processing of the raw material within the
     apparatus and during the extraction process;

b)   production of a fodder meal (meat, blood, feather or bone meal)
     having a high digestible protein content and a good odour, the
     residual fat and residual moisture content being kept below
     4 - 10 per cent;

c)   the production of fats having high saponification properties and
     good bleaching properties and free from unpleasant odours;

d)   good profitability with short charging times;

e)   cleanliness inside the equipment area;

f)   economical in respect of personnel because only valves and
     switches have to be operated.

B. Fundamentals of the process

Perchloroethylene (per for short) is used as a solvent for performing
the wet extraction process. Pure per is a water-clear non-combustible
liquid having a boiling point of 11.95^0C, a specific gravity of 1.62
and a heat of vaporisation of 51.5 kcal. With water or water-containing substances it forms an azeotropic mixture which has a boiling
point of about 85^0C in the case of a per:water ratio of 3:1. This
azeotrope is the basis for the entire wet extraction process and its
low boiling point, which is far below that of pure water, enables the
raw material to be treated extremely carefully resulting in the
above-mentioned high digestible protein contents of the meat meal.
The high fat-solubility of per enables the meat meal to be de-fatted
to less than 4 per cent, thus giving a meal which is storable for
long periods even in a hot climate. Since the first fat removal stage
is carried out with the raw material in the water-saturated state and
the miscella is always heavier than the water-containing material
being extracted, it is possible to process even fermented and fine-

fibred substances which are difficult to extract - if at all possible - with other solvent processes.

C. Operation (see fig 14)

All the equipment required for performing the wet extraction process is illustrated in the following diagram.

1. Comminution of raw material and filling of extractor

The raw material should be so comminuted that it can be introduced without difficulty through the filling nozzle, which projects into the slaughtering chamber, ie 'the unclean side'. In the case of carcasses, it is sufficient for pieces to be comminuted to a size of 40 x 40 cm. Bones should normally be pre-crushed by a bone crusher. In modern large-scale plants, to ensure rapid filling, continuously operating comminuting equipment should be provided with appropriate transport and filling means. Depending on local conditions, the raw material charging can be carried out manually by means of filling tanks or fully mechanically by a preceding crusher, compressed air gun or special pump. The crusher used for this purpose can comminute undivided large carcasses with skin.

When the raw material has been introduced into the combined boiler, dryer and extractor (3) with the agitator in operation (hinged lid or hydraulically or electrically operated valve), the cover is closed over the inlet nozzle (2). Where the special pump is used, the filling cover remains closed.

2. Sterilisation and first drying

The raw material is heated for a short period in the extractor by indirect heating at about 130°C with continuous agitation, and is thus satisfactorily sterilised. The agitator itself is so strongly constructed that the raw material is comminuted and the

y

FIGURE 14

HEB wet extraction process

1   Filler truck
2   Hopper
3   Extractor
4   Condensation plant
5   Miscella tank
6   Filter
7   Distillation apparatus
8   Solvent tank with separator
9   Fat clarification tank
10 Cooling water — inlet
11 Cooling water — outlet

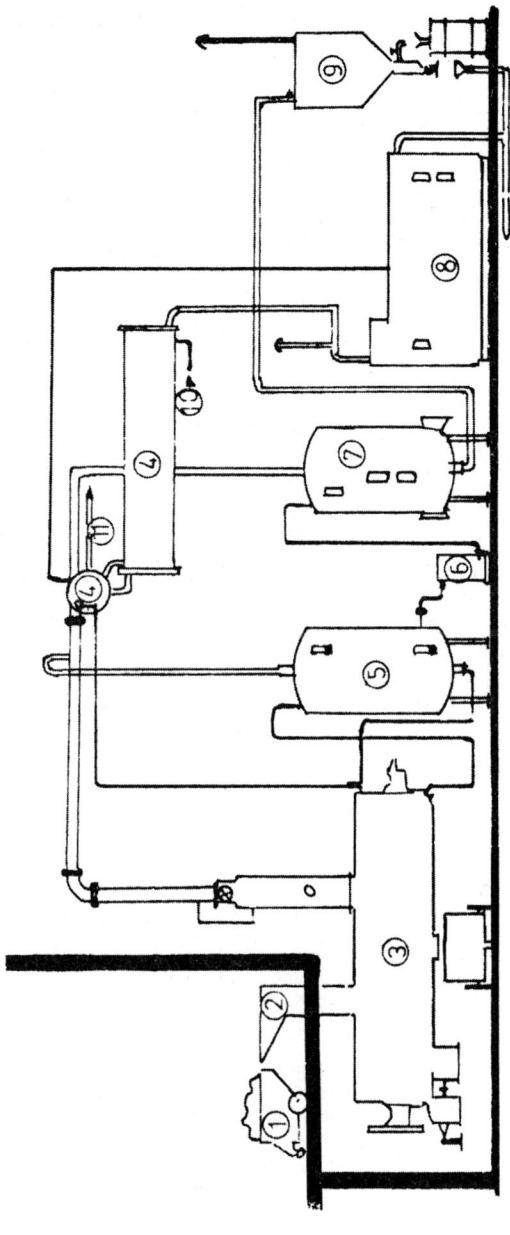

Wet extraction

Source: (18)

soft parts disintegrated during sterilisation. On completion of
the sterilisation process the extractor is depressurised by a
valve above the condensation plant (4), some of the water contained in the raw material evaporating. A subsequent short drying
period expels additional moisture from the raw material. The
actual wet extraction then begins. Where large pieces of bone
are to be processed, and particularly in the case where the charge
is made up only of bones, sterilisation of the material is carried
out at the end of the process.

3. Extraction by first washing

The solvent is pumped into the extractor from the solvent tank (8)
via the solvent pump, passing through the heat exchanger (4a),
where it is pre-heated to about 80 - 90^{0}C, the extractor vapour
heat being utilised. The second miscella of the preceding charge,
ie a weak solution of fat in pure per, is used continuously
instead of pure per. In this case the miscella from the miscella
tank (5) is forced into the extractor by vapour pressure.

4. Discharge of high-fat-content miscella

The miscella highly saturated with fat in the extractor is forced
to the miscella tank (5) after a short period in the extractor
and then via the miscella filter (6) to the distilling unit (7).

5. Second wash and final drying

The per forms an azeotropic mixture in the extractor with the
moisture still present and this mixture vaporises at 85^{0}C. The
azeotropic mixture proportions of per and water are maintained
by continuous replenishment with fresh per from the solvent tank
(8) until the last of the water has evaporated. The end of the
drying operation is indicated optically and acoustically to the
operators by appropriate means. Replenishment with fresh per
also acts as a second wash for further reduction of the fat

content of the raw material. The resulting slightly saturated
miscella is forced into the miscella tank (5) to be used for the
first wash in the next charge.

6. Vaporisation of solvent

The meat meal is already practically anhydrous but which is still
impregnated with per is then subjected to the injection of direct
steam into the extractor with continuous movement of the agitator
to remove the per. In the case of bone processing, sterilisation
takes place at this point by increasing the pressure in the
interior. After the total process has taken some three to four
hours - the time depends on the composition of the charge - the
dry meat meal from which the fat has been removed can be emptied
through the spigot at the extractor (3).

7. Condensation

The vapours forming during extraction - a mixture of water and
per vapour - are precipitated in the condensation plant (4) and
are separated in the solvent and water separator, which is combined with the solvent tank, separation being effected by the
difference between the specific gravities of the two liquids.
The water, which is completely clear, can be discharged to the
sewers while the per is fed to the extractor via the pump.

8. Distillation

The miscella from the first wash, which is highly saturated with
fat, is forced from the miscella tank (5) via filter (6) to the
distilling unit (7) and can already be distilled during the
extraction process. In standard plants, heat is supplied by
indirect steam via the heating register and by direct steam via
a steam jet at the bottom of the distilling unit. Where vacuum
distillation is used, which enables the fat to be treated without
adverse effects as a result of the much lower working temperatures,

the heating register can be heated by the vapours leaving the extractor, thus giving steam savings. However, vacuum distillation requires a vacuum pump and a vacuum receiver, which are not shown in the accompanying diagram.

The vapours forming on distillation are also precipitated in the condensation plant (4) and the condensate is separated in the solvent tank/water separator.

9. Fat clarification

The distilled fat is forced by slight superatmospheric pressure by means of steam into the fat clarification tank (9), where any entrained meal or meat fibre particles can settle. To this end, this vessel is provided with a heating coil and a hot water connexion. The sediments are discharged in a small sludge tank and forced back into the extractor by steam. Depending upon the size of the undertaking, the use of a separator is recommended instead of a number of clarification tanks.

D. Operation of plant

Steam, electric power, cooling water and per are the main requirements for operation for the HEB wet extraction process.

I. Steam

Depending upon the kind of raw material and the composition of the charge, steam consumption has been found to be between 1400 and 1800 kg per tonne of raw material. Depending upon the composition of the raw material and the quantity introduced, the processing time for a charge in the case of bones is about three to four hours and in the case of waste about four to six hours.

2. Electrical power

20 to 25 kWh per tonne of raw material are required for the actual extraction plant. This does not include pre-crushing devices, possible vacuum distillation and the transportation system for carrying off the finished meal.

3. Cooling water

About 20 cubic metres of cooling water are required per tonne of raw material, taking as a basis a cooling water input temperature of 15^0C and an output temperature of 55^0C.

4. Solvent

The per consumption is about 10 litres - 16 kg per tonne of raw material in practice. Given careful monitoring and operation, the per consumption is on the annual average 6 - 8 litres per tonne of raw material.

5. Operating staff

Only one operative is required to operate a HEB plant with two to three extractors.

Procedure in a processing plant for carcasses and slaughterhouse waste operating by the dry extraction process

Technology

There is continuous fat removal by solvent extraction. The entire process is fully mechanised, takes place in a closed system and is divided up into a number of successive stages. This ensures good supervision and reliable operational control via the material flow with 100 per cent recovery of all the fat substances in the raw material, under optimum technological conditions.

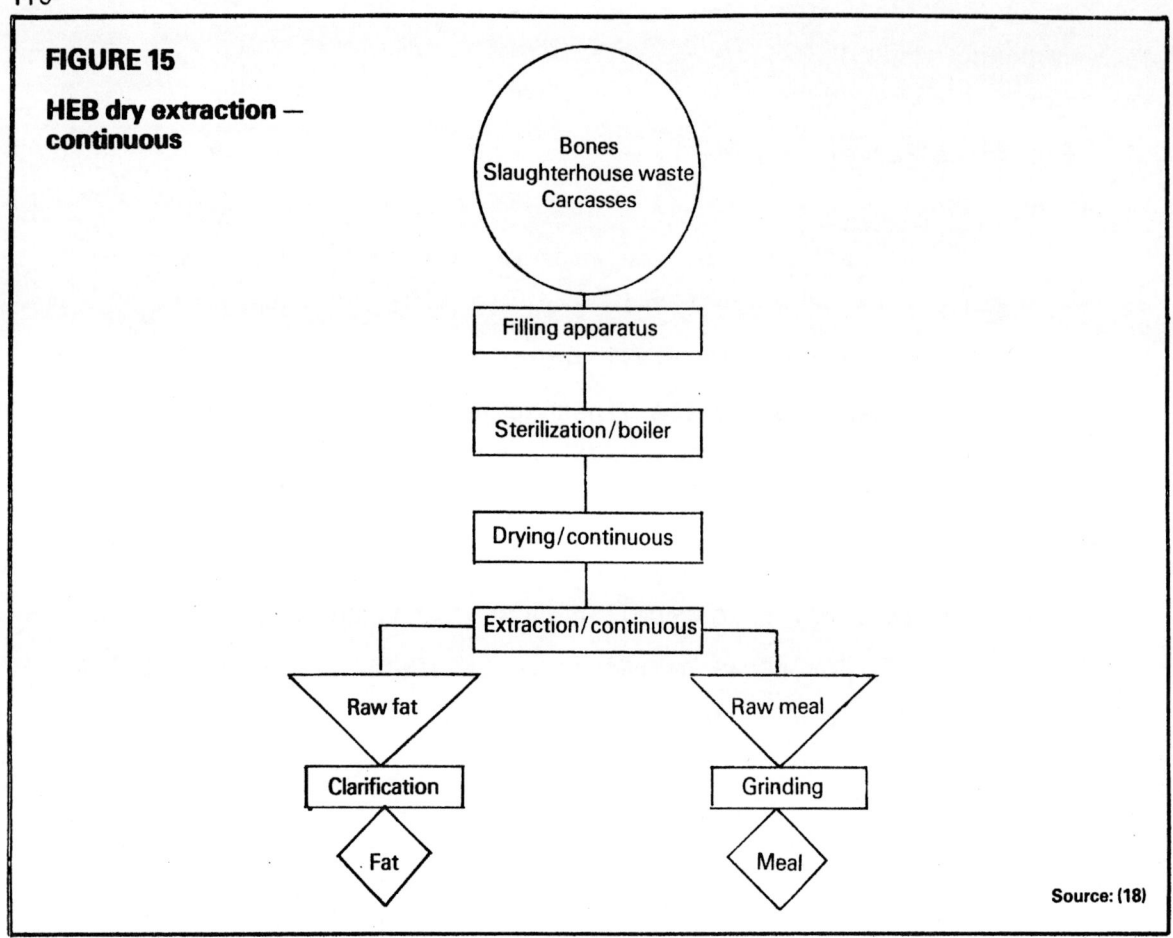

FIGURE 15

**HEB dry extraction —
continuous**

Bones
Slaughterhouse waste
Carcasses

Filling apparatus

Sterilization/boiler

Drying/continuous

Extraction/continuous

Raw fat

Raw meal

Clarification

Grinding

Fat

Meal

Source: (18)

Disintegration and sterilisation (see fig 16)

The raw material is emptied from special vehicles directly into a
large trough (1), which can be covered. A feed screw rotating
therein fills a strainer boiler (2), in which the raw material is
thermally comminuted, disintegrated, and sterilised to form a
homogeneous flowable meat pulp which, after completion of the process, is stored in large-capacity tanks (3).

Drying

The sterilised material is dried indirectly in periodically or
continuously operating agitator dryers (6).

Condensation of the vapours forming during the process takes place
in surface condensers (8); vacuum units (10) ensure careful thermal
treatment of the material for drying. After completion of the
drying process, the intermediate product, from which the fat has not
yet been removed, passes via a closed storage and metering bunker
(7) to the fat removal plant.

FIGURE 16   HEB dry extraction

Sterilization

1 Raw material trough
2 Strainer boiler
3 Intermediate tank for boiled material
4 Cyclone separator
5 Surface condenser

Drying

6 Dryer
7 Bunker
8 Surface condenser
9 Vapour collector tank
10 Vacuum plant

Extraction

11 Extractor
12 Miscella pump
13 Decanter
14 Distillation apparatus
15 Surface condenser
16 Stripping apparatus
17 Surface condenser
18 Solvent collecting tank
19 Solvent pump
20 Dephlegmator

Fat station

21 Fat collecting tank
22 Fat metering pump
23 Separator
24 Sludge recycling apparatus

Meal station

25 Metal separator
26 Hammer mill

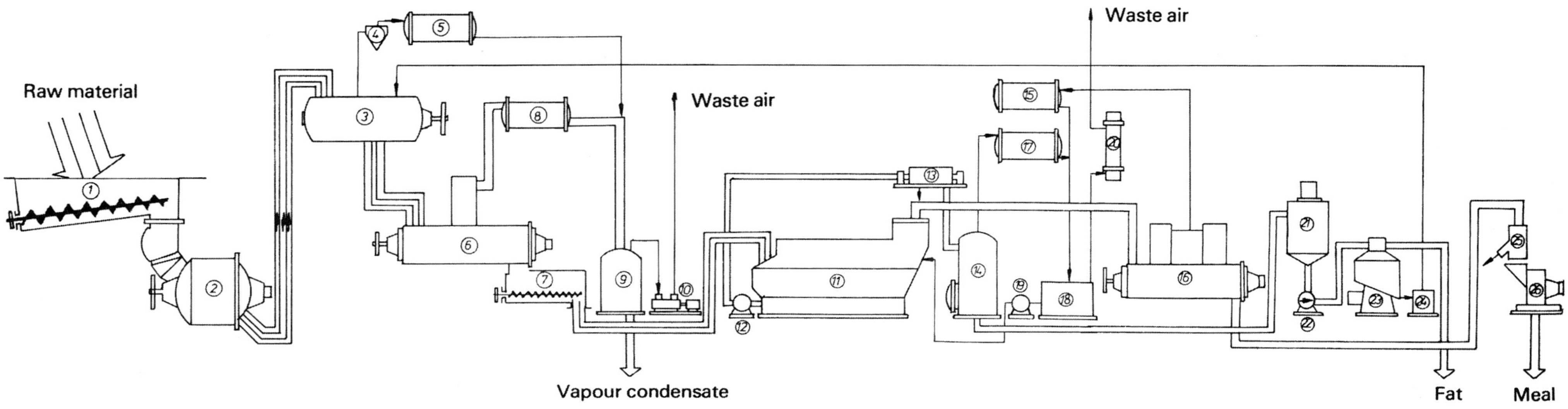

111

Fat extraction

The fat is removed from the material in a fully continuous patented multi-chamber extractor (11) by means of solvents in counter-current under optimum economic conditions until there is a specific residual fat content. The end products - animal meal and industrial fat - leave the extraction plant free of solvents and pass mechanically or pneumatically to the next processing plant.

Processing of end products

Meal: After passing through a metal separator (25), the meal is ground in a mill (26) and packed in bags ready for dispatch or stored in bunkers.

Fat : The fat obtained in the extraction unit is processed in a fat clarification station (21 - 24) and then pumped into fat tanks.

Odour elimination

All the processing chambers contain air ducts. A change of air at a fixed rate per unit of time copes with all the odiferous outgoing air, which is disposed of in a patented deodorising plant.

Complete destruction of the non-condensable gases from the condensers and vapour tanks is carried out in a heavy-duty incinerator plant. This system avoids any smell nuisance in the plant and in the immediate and further surroundings of the carcass disposal institution.

Procedure for bone processing

Contrary to the above plant design, for bone processing items 2 - 5 (boiler and tank with accessories) are replaced by a crusher and a feed unit. Periodic sterilisation takes place in the unit (6) after drying. The intermediate bunker (7) is replaced by a screening unit which is intended to separate the fines from the fat-containing dry

substance and feed all the particles larger than the required size to
a crusher. This ensures careful comminution with the smallest fines
fraction. The technology is particularly important in respect of optimum particle size of the semi-product for the glue and gelatin industries.

10.4   COST ANALYSIS OF THE VARIOUS DISPOSAL AND PROCESSING METHODS

10.4.1 *Waste water*

The processing costs for waste water differ considerably depending on
the size and degree of utilisation of the purification plant.

10.4.2 *Costs of disposal and processing of waste by animal carcass disposal institutions*

Processing costs differ according to production process, size of undertaking and capacity utilisation. They are in the range from DM 30 to
DM 150 per tonne of raw material. Transport costs depend upon the
catchment area. Generally they are between DM 50 and DM 70 per tonne
of raw material.

The costs for the clarification or deodorising plant for waste water
are between DM 5 and DM 10 per tonne pf raw material.

Example 1

A 'I x 2000' wet extraction plant made by Krupp will be given as an
example.

The capacity of this plant is 10 t raw material/24 hours.

Yield: meal = 23% = 2.3 t
        fat   = 12% = 1.2 t

Prices (May 1974)

Meal   DM 360/t

Fat   DM 460/t

For a yield of:

Meal 2.3 t x DM 360 =   DM   828

Fat 1.2 t x DM 460 =   DM   552

Net proceeds   DM 1380

Less transport costs (about DM 60 per t)

10 t x DM 60 =   DM   600

Less waste water costs

(about DM 8 per t)

10 t x DM 8 =   DM   80

Less wage costs

24 hours x DM 15 =   DM   360

Less cost of plant

(about DM 70 per t)

10 t x DM 70 =   DM   700

Total costs   DM 1740

Example 2: HEB pressing process (dry process)

Capacity 60 to 80 t/24 hours

Average 70 t/24 hours

Yield:

Meal 23% = 16.1 t

Fat 12% =   8.4 t

Prices as in the preceding example.

For a yield of:

Meal	16.1 t x DM 360 =	DM	5796
Fat	8.4 t x DM 460 =	DM	3864
Net proceeds	=	DM	9660

Less transport costs

70 t x DM 60 =   DM   4200

Less waste water costs

70 t x DM 8 =   DM   560

Less labour costs

24 hours x DM 15 =   DM   360

Less cost of the plant

70 t x DM 35 =   DM   2450

Total costs   DM   7570

The costs per tonne of raw material are as follows:

Wet extraction process	DM	174/t
Pressing process	DM	108/t

Waste can therefore generally be processed more cheaply by the pressing process.

These examples clearly show that transport costs make up 55 per cent of the costs on average.

To make a plant economic and profitable, the transport costs must be reduced to a minimum.

The consequence of this would be centralisation of the meat-processing industry in conjunction with the waste-processing industry.

The costs listed do not include capital and building costs.

10.5 COST BENEFIT AND COST LOSS ANALYSES IN CONNEXION WITH THE DISPOSAL OF SLAUGHTERHOUSE WASTE

The exponential world population growth has greatly restricted the space available for the individual. The danger to health grows increasingly unless appropriate waste disposal measures are taken.

The denser the population of a given area, the greater the emphasis on safeguarding health, which becomes the first commandment.

The disposal of all problematic waste, which includes slaughterhouse waste, must thus be carried out in the most productive manner, ie perfectly from the veterinary aspect. The cost principle is secondary in this context, because the main emphasis is on the hygienic aspect. The economic and national significance in respect of profitability and food policy take second place. If slaughterhouse waste is processed, it constitutes a considerable advantage to the national economy as protein food.

If we consider disposal from hygienic and economic aspects, then the following disposal systems come into consideration in sequence:

1. Utilisation for human nutrition (possible only to a restricted degree)

2. Special processing from aspects which are satisfactory hygienically and in respect of veterinary hygiene

3. Utilisation of problematic waste (condemned meat, etc) via animal carcass disposal institutions

4. Incineration of poisoned carcasses.

Only the hygienic and environmental aspects are concerned in the case of the disposal of noxious substances in waste water. In principle, no European country can afford ultimately to disregard the hygienic aspects and place all the emphasis only on the economic aspects. The extreme case that comes to mind is the burying of problematic waste.

In principle, the costs incurred in disposal do not play a dominant part in this study, because the loss that may arise due to the national economy as a result of contamination and disease resulting from improper disposal is very much higher than the expenditure required to dispose of problematic waste from aspects which are satisfactory in respect of contamination and veterinary hygiene.

The benefit that can be obtained by processing waste water sludge together with domestic refuse by composting, is also very important. The increasing use of technology and chemistry in agricultural production gives rise to problems which grow disproportionately with increasing use of agricultural aids such as mineral fertilisers and plant protection agents.

As a result of the increasing use of these agents, a fall-off has been found in the productivity of agricultural soils. To obviate this diminishing productivity, mineral fertilisers are increasingly used as a protective agent. The result of these abnormal applications has in some cases been devitalisation of agricultural soil. This devitalisation is progressive, so that in the foreseeable future the once vitalised soil can be used only as a plant station. Plants which grow on these soils have to be positively synthetically nurtured, because the natural nutrient supply of the soil is no longer fully guaranteed for the reasons given above. The excessive use of substitute fertilisers and protective agents does not guarantee qualitative growth. Only mass production can be achieved and this is coupled with disproportionate degradation of the organic substance in the soil.

Plants grown in this way are subject to a deficient supply of trace elements and nutrients and are therefore particularly susceptible to plant diseases. To be able to counteract these plant diseases, highly active and often toxic agents are used mainly symptomatically for treatment, owing to the lack of any other possibilities.

The products produced in this way cannot be regarded as high-grade in either human or animal nutrition, because the internal quality does not satisfy the natural requirements for healthy nutrition even in the presence of sufficient external mass. The residues from fertilisers and plant protection agents may produce diseases in man and beast.

The soil is a living organism. It contains dissolved and undissolved mineral constituents, water, organic substances which are deposited from living plants, and organic and inorganic substances arising from the decomposition of roots and plants. The soil is interspersed with living creatures, such as earthworms, bacteria, insects, larvae, and higher animals, the activity and decomposition of which contribute to decomposition of the soil, physically by loosening it, and chemically by absorption and digestion. The co-operation of all these factors with the climate and human cultivation determine the soil fertility.

Soil fertility and hence plant growth are influenced by the organic substance content of the soil.

Sufficient organic substance in the soil provides favourable living conditions for the soil flora and fauna. The organic substance content of the soil is termed humus, but not all the organic constituents of the soil need necessarily be present in the form of humus. Some 40 per cent of the organic constituents are substances similar to humus, the exact chemical formula of which has not yet been established.

It is a mixture of degradation products rich in carbon, nitrogen and oxygen and differing in synthesis and degradation in each individual case. The condition of the humus is an important factor. There are three main processes taking place in the organic structure of the soil:

1. the degradation of the raw materials used to form humus,
2. the degradation of stable humus substances, and
3. the maintenance or loss of synthesis substances.

If there is equilibrium between these three processes, the soil may
be termed healthy but if the equilibrium is disturbed the soil is
sick and suffers, *inter alia,* from metabolic disorders.

It is thus possible to establish that with humus contents below 1.5
per cent the function and progress of soil life are imperilled and
below this limit the function of the soil life is considerably disturbed. The productivity of soils for cultivation can be maintained
permanently only by supplying organic substances, the maintenance of
the stock of humus in the soil being a decisive factor.

In this connexion reference should be made to the problems arising
from the fact that, for example, in the Federal Republic of Germany
the organic waste substances in solid and liquid form are not returned
to the natural cycle - apart from negligible exceptions - and instead
are disposed of in incinerators or on controlled and uncontrolled
dumps where, in most cases, they pollute the environment in the long
or short term.

This statement applies not only to the Federal Republic of Germany;
many neighbouring states within the EEC suffer from the same problem.

This short description is intended to show the advantage that can
arise if liquid waste, not only slaughterhouse waste, but also
communal waste, are composted together with domestic refuse in a
composting plant and fed to the soil as organic substances.

Because of the numerous inter-relationships between waste disposal
and processing, the economics thereof, and their effects on health
and hygienic conditions, it is not possible to compile data as to the
costs or profit arising from waste disposal and use.

10.6 ASSESSMENT OF THE DIFFERENT DISPOSAL AND PROCESSING METHODS

The disposal methods are:

1. Waste water
2. Special processing
3. Carcass disposal institutions
4. Disposal by heat.

1. Waste water

Care should be applied during the actual slaughtering process to ensure that the minimum amount of organic substances pass to the waste water system.

The processing method by means of sludge digestion and composting is very good. Against this, thermal processing or disposal of waste water sludge is undesirable, because the costs are too high. Incineration also simply means a reduction of the quantities of waste accumulating.

2. Special processing

It may be assumed that special processing is economic, because the industries engaged therein will undertake processing only if it is economic. Otherwise these quantities of waste also go to the carcass disposal institutions.

3. Carcass disposal institutions

The processing methods used in these institutions are satisfactory from the veterinary aspect, but deodorisation of the waste water and waste air is an urgent matter. The problem of the economy of these disposal methods does not lie in the technical area so much as the structural area (see 10.4). This structural problem applies not only to the carcass processing institutions, but also to their links with the slaughterhouses and processors. Given an appropriate correlation between sufficient quantities of waste and sufficient utilisation, the slaughterhouse waste can be disposed of economically. The carcass disposal institutions also produce considerable quantities of protein reserves for animal feeding purposes.

4. *Thermal disposal (incineration)*

Hygienically, the thermal process is unobjectionable. Economically, however, considerable quantities of energy have to be supplied, while the efficiency of the resulting quantities of heat is minimal, so that this method should be applied only for the disposal of poisoned animals.

10.7 POSSIBLE USES OF SLAUGHTERHOUSE WASTE

In this connexion see more particularly 10.1.

1. *Possible uses in industry:*
Leather goods, brushes and paint brushes, buttons, fire extinguishers, pet foods, glue, gelatin, photographic, and pharmaceutical industries, the chemical industry and shoe polish industry and so on.

2. *In agriculture:*
The main uses in agriculture for slaughterhouse waste are by way of carcass meal, meat and bone meal, coarse bone meal, feeding bone meal, bone ash, bone charcoal, calcium phosphate and bone fat for animal feeding purposes.

Poultry waste can be used directly for animal feeds if the sterilisation regulations are complied with, but use of this kind is not desirable.

11 Economic significance of the utilisation of slaughterhouse waste

This has already been discussed in other sections, and here we shall point out that the utilisation of slaughterhouse waste via the various disposal systems can be said to be of considerable economic benefit to the national economy, subject to the production of disposal plants.

The economic significance of slaughterhouse waste to agriculture lies primarily in the fact that considerable quantities of protein reserves can be produced for animal feeding purposes.

Certain slaughterhouse waste which is not condemned can be used for the production of organic fertilisers. This includes particularly skin, bristles, claws, horns and stomach and intestine contents.

The manure from the slaughterhouse can also be used as an organic fertiliser in agriculture. These quantities are economically insignificant.

With regard to the use of waste from slaughtering as an organic fertiliser, however, reference should be made to the secondary infection possibility, even if such material primarily contains no pathogens, since substances having a high organic content provide good possibilities for the proliferation of pathogens.

Generally speaking, the use of slaughterhouse waste (of the kind suitable for organic fertilisation) in agriculture does give rise to advantages over simple disposal, since the latter method entails only costs without any benefit.

12 Significance of the utilisation of slaughterhouse waste to the environment

This section will explain the importance of the utilisation of slaughterhouse waste to the environment, ie to man's living conditions.

Not only is utilisation relevant to maintaining environmental health, but in addition disposal should be carried out by hygienically perfect methods. The fact that the economics of the problem play a large part will be mentioned by the way. If disposal were carried out only from economic aspects, there would be the danger of considerable pollution of the environment by pathogens and contaminants, which are a result of the rapid putrefaction of this material.

Slaughterhouse manure can be regarded as hygienically perfect, subject to prior stacking and storage. If the manure is stacked and stored for a relatively long period, spontaneous heating occurs during such period and may reach temperatures of up to 70^0C, sufficient to kill the contaminants and pathogens (bacteria, viruses or parasites).

Reference has hitherto been made only to pollution of the environment by pathogens. Another extraordinary pollution resides in a great danger of smell nuisance due to the rapid putrefaction of the waste. It is mainly ammonia, hydrogen sulphide, fatty acids and mercaptans (13 + 19) which occur on the deconposition of natural products.

In carcass disposal institutions in which thermal decomposition is mainly carried out, the smell nuisance is produced mainly from aldehydes, ketones, fatty acids and unsaturated compounds. Fumes and vapours are the main vehicles for these substances.

To reduce this smell nuisance to a minimum, it must be taken into account when the raw material is actually delivered, so that the material is processed in the freshest possible state. If it is not possible to transport the raw material from its place of origin immediately, then the raw material must be chilled during storage.

The material should be transported only in closed or covered containers complying with general hygienic and veterinary requirements.

Carcass disposal institutions should be so designed as to be separated into a clean side and an unclean side. The raw material should be processed in a closed system so that the smell-bearing substances can be suction-extracted and disposed of harmlessly.

VDI Code of Practice No. 2590 contains a number of technical possibilities for air purification[13]

These Codes of Practice distinguish between two possibilities:

1. reduction of the emission of smell-intensive substances;
2. waste air purification processes.

With regard to the possibilities of reducing the emission of smell-intensive substances, a distinction is made between:

A. organisational action,
B. technical action.

A. Organisational action

By organisation is meant that the raw material should be rapidly collected on the same day as it arises, in order to avoid smell emission. The raw material should also be chilled until it is collected. The capacity of the processing plants should be so designed that all the raw material can be processed without delay.

B. Technical action

The technical action provides for a closed apparatus system.

The waste air cleaning processes are as follows:

1. The wet process
   a) limestone tower process
   b) multi-stage scrubbing process
   c) potassium permanganate process.

2. Dry processes (incineration)
   a) Thermal post-incineration
   b) Catalytic post-incineration

3. Biological processes

12.1 PARTICULAR PROBLEMS IN CONNEXION WITH COMPOSTING

Slaughterhouse waste falls into different categories in respect of
composting. In principle, all waste occurring at the slaughterhouse
can be composted. The waste from waste water (clarified sludge)
provides no difficulty if it has been pre-clarified and the waste
from animal excrement (manure) does not give rise to any difficulty.
The clarified sludge can be composted together with domestic refuse
or by the 'compostabilizing' method[16, 20] without any special processing. Composting in stacks (Fahr AG System[2]) is particularly
suitable for composting animal excrement in conjunction with domestic
refuse and clarified sludge. The temperatures obtained in composting
are important to its perfect performance. To obtain hygienically
perfect compost, temperatures of more than 55°C must be achieved
during composting. Table 32 shows the hygienic quality of the
individual composting processes according to Knoll[22] and Strauch[21].

Table 32: Hygienic assessment of composting processes (according to Knoll and Strauch[21])

Method	Material	Water content %	Maximum temperature achieved	Duration	Hygienic assessment	Notes
Open composting						
'Cold composting' in flat stacks	Domestic refuse – clarified sludge	55	46°C	5 months	Not perfect	
	Clarified sludge	60	52°C	6 months	Not perfect	
'Cold composting' in stacks	Domestic refuse	40–60	55°C	3 weeks	Perfect	turn over once
Stack composting	Refuse and clarified sludge	40–60	55°C	3 weeks	Perfect	turn over once
Stack composting	Refuse and clarified sludge	40–60	55°C	3 weeks	Perfect	turn over once
Fahr AG system	Refuse and animal excrement					
Surface composting	Refuse and clarified sludge – hygienic investigations not yet completed					
System composting						
Movable rotting cells						
Rotary drum	Refuse	45–56	60°C	6–7 days	Perfect	for spore-formers Plus 4 days stack composting
(eg Dano process) Rotary drum	Refuse and clarified sludge	about 50	about 60°C	6–7 days	Perfect	
Rotary drum with aeration and venting *) (RK process Bühler-Rheinstahl)	Refuse and clarified sludge	45–55	67°C	3 days	Perfect	for spore-formers + 3 days stack
Rotting tower with central rotary axis (eg Multibacto process)	Refuse	40–50	65°C	1 day	Perfect	for spore-formers + 6 days stacks + 6 days stacks

*) see figure 17

129

Table 32: continued

Method	Material	Water content %	Maximum temperature achieved	Duration	Hygienic assessment	Notes
Rotting tower	Refuse and clarified sludge	45–55	65°C	1 day	Perfect	
Stationary rotting cells with aeration and circulation (eg Diefenbacher)	Refuse and clarified sludge	about 50	56°C	5 days	Not perfect	
Stationary rotting cells with aeration – Blaubeuren system	Refuse and clarified sludge	35–50	70°C	12–14 days	Perfect	
Other processes Capillary drying process (eg Brikollare process) Cast model	Refuse and clarified sludge	40–55	60°C	3 weeks	Perfect	
(Modified Voith-Müllex system rotting dump,	Refuse and bulky refuse	45–50	70°C	21 days	Perfect	
	Clarified sludge		60–70°C	3–5 weeks	Perfect	
Aerated dump	Refuse and clarified sludge	45–55	70°C	3 weeks	Perfect	

Strauch and Knoll distinguish three processes (see table 32):

1. Open composting
2. System composting
3. Other processes

System composting is divided into:

a) dynamic rotting cells
b) static rotting cells

The advantages of system composting are that a hygienically perfect fresh compost is obtained after a relatively short time (one to fourteen days depending upon the system). To reach full maturity, this fresh compost can be stored on a stack for a few days or weeks.

The composting of slaughterhouse waste in the true sense as defined in section 2 gives rise to certain technical problems at the present time.

These are due to the fact that the slaughterhouse waste must be comminuted before composting. The resulting meat pulp stops the further transport of the rotting material in the rotting cells. If slaughtering waste is composted in stacks, longer rotting times are required. If the proportion of slaughterhouse waste is too high, turning over of the stacks becomes a problem. If the stack is not turned over, there is the risk of an anaerobic process (digestion process) occurring instead of the aerobic process.

The hygienic aspect must not be disregarded in connexion with composting.

Industry has taken part only marginally in the composting of slaughtering waste because at the present time there is no demand for composting plants in which slaughtering waste can be composted.

FIGURE 17

Diagram of an RK composting plant

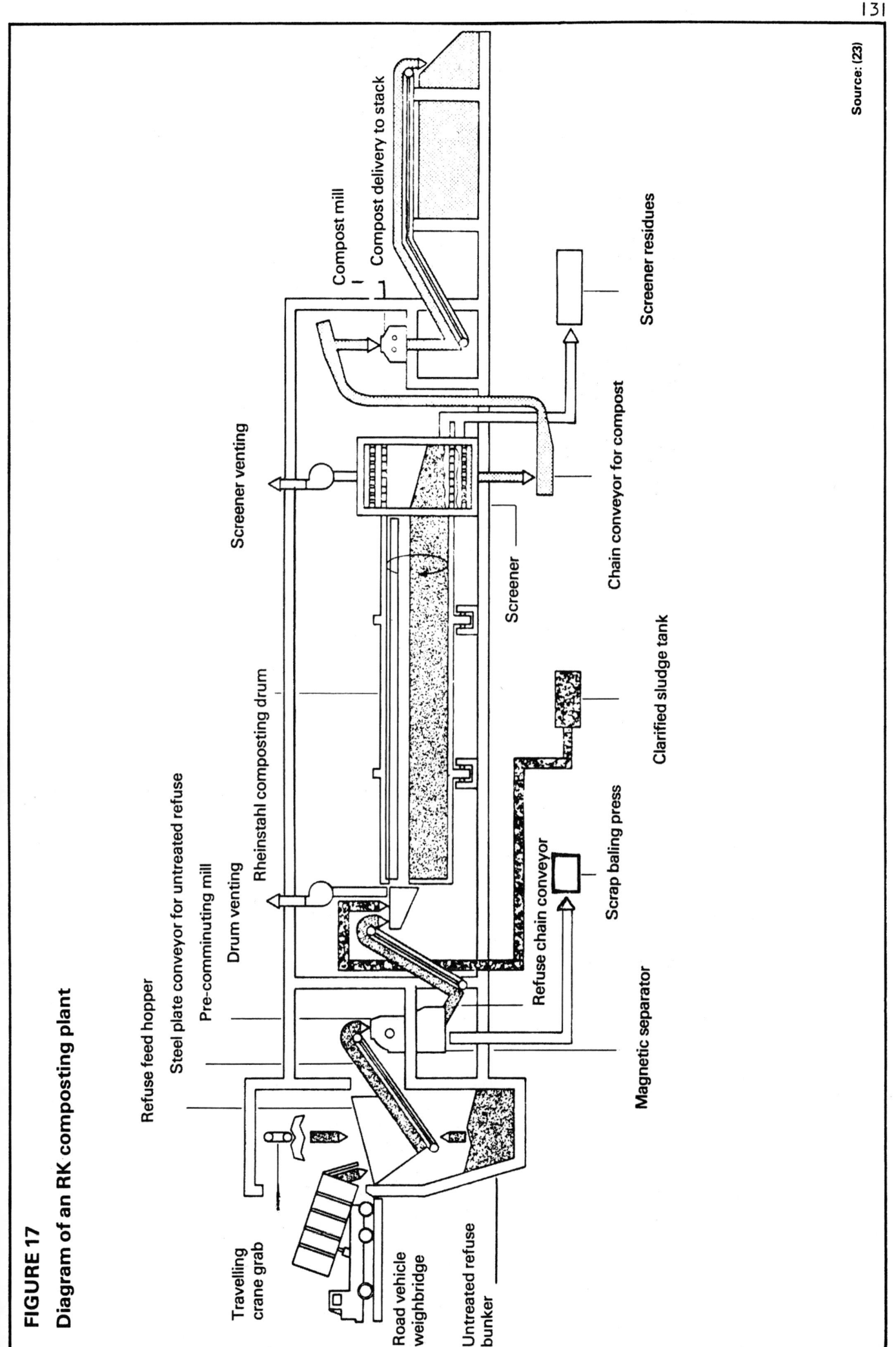

Source: (23)

13 Proposals for the future disposal of slaughterhouse waste in the EEC (ideal situation) and for the use and utilisation of slaughterhouse waste from economic and environmental aspects in industry and agriculture

13.1 IDEAL SITUATION IN THE EEC

As shown in fig 18, the ideal situation in the EEC would appear as follows:

1. Waste which cannot be used for human nutrition because it was hitherto refused by the consumer because of consumption practices, should be used for human consumption via new marketing channels.

2. Waste which can be processed by special processing should as far as possible be collected and processed by the industries concerned. To make such utilisation economic, the numbers of animals killed at any one place should be high.

3. Waste which is condemned and waste which cannot be processed via the above processing channels should be collected and processed by carcass disposal institutions.

   The products made therefrom can on the one hand be sold via industry and, on the other hand, via agriculture and be economically used by them as described above.

4. Waste which is disposed of via the sewage system must be recovered by prepurification before being discharged into the sewers.

5. The carcass disposal institutions must also accept and dispose harmlessly of waste which occurs in the form of dead animals.

As shown in figure 18, all the areas within a district should be subjected to certain statutory regulations to permit only disposal systems and methods which are acceptable from veterinary and public health aspects.

134

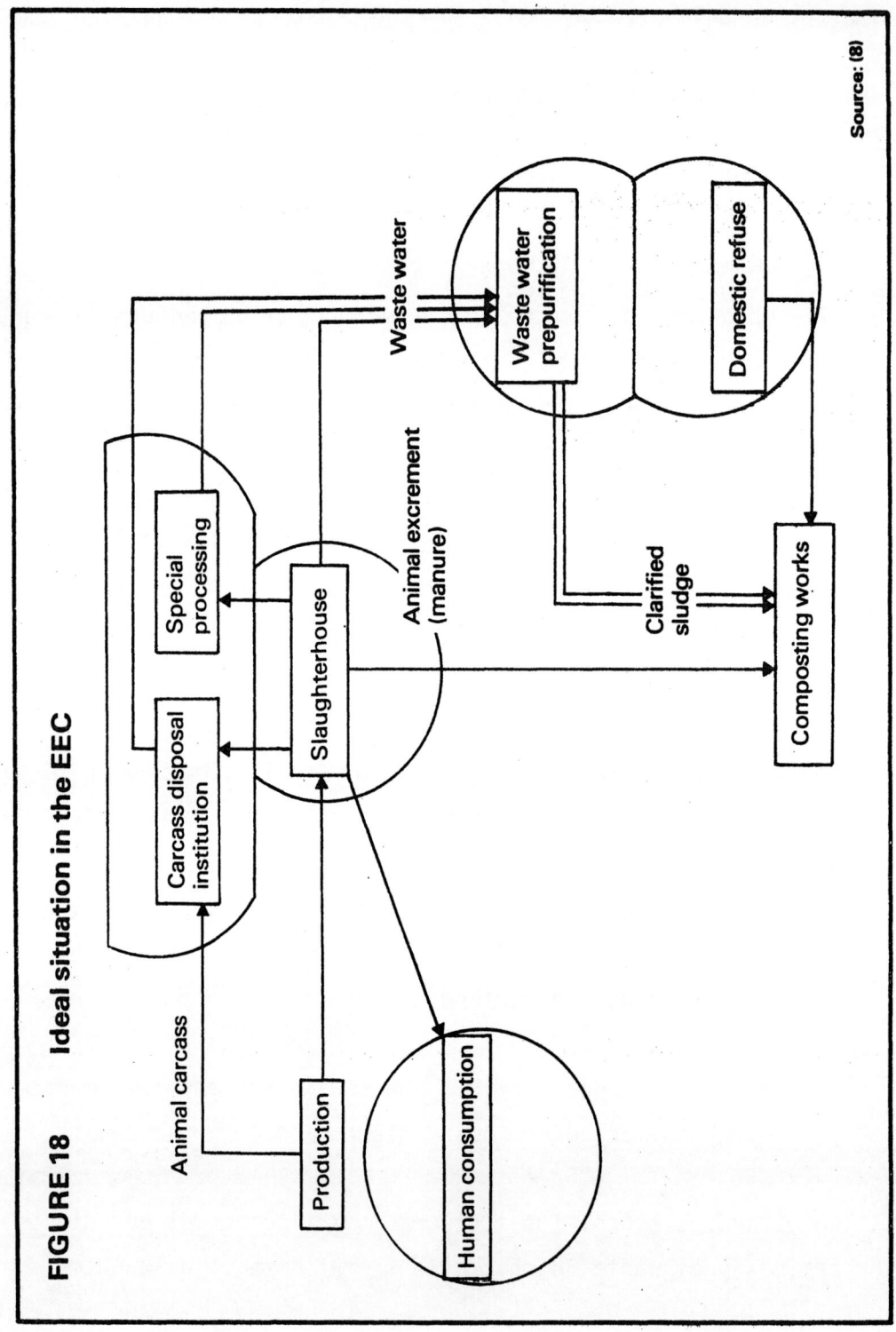

FIGURE 18   Ideal situation in the EEC

Source: (8)

Under the heading of special processing, slaughterhouse waste is processed solely in the private industrial sector, which undertakes such processing only if it is economic. For this sector, the statutory regulations should be confined to ensuring that processing is carried out under hygienic aspects.

In addition, care must be taken to ensure that if this waste does not undergo special processing it is collected and disposed of by carcass disposal institutions.

The legislation in respect of the carcass disposal institutions should be extended to make them responsible for hygienically perfect processing of problematic waste; they must also be statutorily bound to dispose harmlessly of all dead animals occurring in their regional area.

Administrative regulations should give each carcass disposal institution a catchment area which as far as possible guarantees economic processing under hygienic conditions.

A condition to this is the drafting of carcass disposal plans to stipulate the structure of slaughterhouses, carcass disposal institutions and agricultural undertakings.

In purely agricultural areas in which dead animals occur and only small quantities of slaughterhouse waste, disposal via the carcass disposal institutions must be guaranteed.

A carcass disposal institution which has to carry out hygienic disposal with small quantities of raw material and very high transport costs is not able to operate economically, as will be apparent from the calculated examples in 10.4.2.

In such cases the State must provide subsidies to enable proper operation to be maintained.

The obligation as to disposal should be transferred by the State to private institutions so that private enterprise is taken into account and the State is thus relieved of the burden. The State should only provide supervision in respect of the statutory regulations.

The statutory regulations should appear as follows:

Each processing or disposal area must be governed by legislation, figure 18 shows these areas by circles.

The following laws would be required for all the EEC Member States:

1. *An Act governing the slaughter of animals*
    This Act should control slaughtering in the light of animal protection provisions to ensure that the animals are killed properly and painlessly.

2. *An Act governing the disposal of animals, carcasses and slaughterhouse waste*
    The scope of this Act should control the harmless disposal of carcasses and parts of carcasses. It should cover the delivery, collection, transportation, storage, burying, incineration, trade in and processing of carcasses and parts thereof and animal products. In the case of special processing, the following should be taken into account and stipulated by legislation:

    1. The health of man and beast must not be endangered by pathogens.

    2. Water, soil and fodder must not be contaminated by pathogens and toxic substances.

    3. The environment must not be polluted by smells, contamination and unpurified waste water.

    No products for human consumption must be produced in these areas.

3. *An Act governing the sale of meat and meat products*

An Act for the inspection of meat must stipulate that animal products intended for human consumption must undergo inspection to ensure that no objectionable meat products are marketed.

4. *An Act governing the disposal of waste water*

The areas of waste water purification of carcass disposal institutions, special processors and waste water from slaughterhouses must be cobtrolled by law. Waste water from undertakings may be discharged into communal sewer systems only after prior purification. The clarified sludge resulting from purification must be disposed of harmlessly.

5. *An Act governing the disposal of waste*

An Act governing the disposal of waste should ensure that the area of refuse disposal is completed by modern and hygienically perfect methods and that the recycling process comes to the fore.

13.2 EXAMINATION OF ALTERNATIVE SOLUTIONS TO SEE WHETHER THEY CAN BE EMBODIED IN VIEW OF LEGAL AND ADMINISTRATIVE REGULATIONS

Many EEC countries already have various laws and regulations. EEC directives or orders should provide generally valid regulations so that the disposal of problematic waste can be carried out uniformly in the EEC area under hygienic conditions which do not result in pollution.

13.3 SUGGESTIONS FOR THE DEVELOPMENT AND ADAPTATION OF LEGAL AND ADMINISTRATIVE REGULATIONS FOR THE DISPOSAL OF SLAUGHTERHOUSE WASTE IN THE EEC

As already stated, specific EEC orders should provide for all the EEC countries to issue Acts and Orders for their area and to harmonise existing Acts and Orders. The top priority is to safeguard the health of man and beast.

14 Summary

Slaughterhouse waste is divided up into slaughtering waste and other waste, the slaughtering waste including waste which cannot be used for human consumption. The other waste includes the waste resulting from the stabling of animals before slaughtering, ie faeces, urine and litter. Slaughterhouse waste is thus made up of slaughtered carcass waste, condemned meat, animal excrement, litter and waste water.

Coefficients specific to each type of animal as specified by Schön, Horn and Belendorff on the basis of considerable investigation are used to determine the amount of slaughtering waste. To express the other waste numerically, estimates have been made in some cases because no sufficient statistical data are available.

Vanselow divides slaughtering waste by type and in proportions by reference to the slaughtered carcass. Since carcass weights differ in the EEC countries, the average carcass weights were determined for the individual EEC Member States in order to determine the quantities of the types of animal and have been taken as the basis. Tables 18 to 26 show the quantities of waste by type and quantity for the individual EEC Member States. The number of animals slaughtered in the individual States has been taken as a basis for this. Poultry waste makes up about 20 per cent of the live weight, 6 per cent being feathers and blood and the rest the other types of waste.

Some of the present-day disposal methods cannot be considered as unobjectionable in respect of contamination. The burying of animals and parts of carcasses should be particularly mentioned. This type of disposal is extremely objectionable from the aspect of hygiene and public health and should therefore be absolutely abjured.

The disposal of slaughterhouse waste by the carcass disposal institutions is unobjectionable, the extraction and pressing processes being particularly noteworthy. This is subject to compliance with sterilisation and hygiene regulations. These processes yield meal for animal fodder and industrial fats for the production of detergents.

The processing of waste by special processing can be regarded as economic because the undertakings which carry out special processing operate only from economic considerations. The carcass disposal institutions, however, are unable to operate solely by economic considerations, because they often have to take waste which is or cannot be processed by the special processors. This means that in some EEC countries the carcass disposal institutions are legally bound to accept this problematic waste and disposal of it harmlessly insofar as concerns the environment. The carcass disposal institutions thus perform a function of value not only to the national economy but also to public health.

Some 25 per cent of the bones go with the meat to the final consumer and are therefore lost to the trade. Only the faeces, urine and litter occurring should be used as organic fertilisers in agriculture. Other organic fertilisers are horn and claws, which are used in the form of horn shavings.

The biogas plants, which produce methane gas from slaughtering waste, have not become established in practice, because there is usually no continuous supply of waste.

Many of the EEC Member States already have a number of Acts and Orders specifying the harmless disposal of slaughterhouse waste. Harmonisation of the Acts and Orders in the EEC area is urgently essential in order to avoid jeopardising public health in the European area.

An improvement in disposal from the hygienic and economic aspects can be obtained only by achieving co-ordination between the slaughterhouse and the waste-processing undertakings.

The future development of slaughterhouse waste depends on various factors:

1. Population growth
2. Per capita consumption
3. Per capita income

It is also dependent upon the numbers of animals slaughtered which in turn depends on supply and demand, which yet again depend on the above three factors.

The number of animals slaughtered increased on average by 100 per cent from 1950 to 1970. A careful estimate of the growth in meat consumption during the next ten years is 30 to 50 per cent; the same applies to the amount of slaughterhouse waste.

New administrative regulations are under discussion in many EEC Member States. This should be a start for co-ordination and harmonisation of regulations.

There are possibilities of using slaughterhouse waste both in industry and in agriculture.

Special waste constitutes particular problems in respect of disposal. As far as poisoned animals are concerned, only incineration applies. Burying should not be carried out for the disposal of special waste either.

These are the most diverse processes for the re-utilisation of slaughterhouse waste, waste water purification and carcass disposal having been discussed in detail in this context.

The important processes for the re-utilisation of slaughterhouse waste in carcass disposal institutions are:

1. Dry process
2. Perchloroethylene process
3. Wet process.

The wet process is little used today because the end products are of lower quality. The pressing process is the most economic of the dry processes. The perchloroethylene processes can be divided up into:

a) wet extraction and
b) dry extraction.

The difference is that water is removed before the extraction operation in the case of dry extraction.

All the processes can be regarded as hygienically perfect if regulations are complied with.

Transport costs are an important factor in the costings of the carcass disposal institution.

These transport costs can greatly influence the economy of a carcass disposal institution. If the institution is in an unfavourable position in relation to the slaughterhouse, transport costs may make up 55 per cent of the total costs.

Since the carcass disposal institutions have played a predominant part in maintaining public health and safeguarding the environment, the economic aspect should take second place.

The composting of slaughterhouse waste still gives rise to considerable technical problems today due to the fact that at the present time there is no demand for composting plants in which slaughterhouse waste alone can be composted.

There are no technical problems in the composting of waste water sludge and waste from stabling (manure, litter, etc). Communal composting plants can take these waste substances together with domestic refuse and process them into compost.

Sections 7 and 8 dealt with the analysis:

    a. of the existing and planned legal provisions, and

    b. the organisation of the disposal of slaughterhouse waste
       in the Member States.

For this purpose, the competent ministries of the individual EEC
Member States were approached and asked to answer important questions
compiled in a questionnaire. In some cases the answers left much to
be desired, while in other cases no answers were received to the
questions by the time this paper was completed.

From this, and from the assessment made in the above Sections, it is
clear that the legal provisions in many of the Member States are
inadequate and the organisation of disposal has not been given an
optimum solution from hygienic and public health aspects.

In order to maintain public health and the environment, legal and
administrative provisions within the EEC should be harmonised.

Bibliography

(1)   Dennis Meadows *Die Grenzen des Wachstums* Deutsche Verlagsan-
        stalt Stuttgart

(2)   L. Schön, K. Holz and M. Belendorff *Ausmass und Minderung von
        Umweltbelastungen durch Verarbeitungsrückstände der Fleisch-
        wirtschaft Bericht über Landw.* Vol. 50 - 1972 pp 675 - 681

(3)   U. Vanselow *Zur Verwertung und Bedeutung der Schlachtneben-
        produkte* Diss. Munich 1970

(4)   Eurostat 7/1973

(5)   SAEG *Unterlagen vom Statistischen Amt der Europäischen
        Gemeinschaften* Luxembourg

(6)   *Statist. Jahrbuch 1973* Bundesministerium für Landwirtschaft
        und Forsten, Paul Parey-Verlag Hamburg

(7)   U. Vanselow 'Schlachtnebenprodukte des Geflugels' *Jahresbericht
        der Bundesanstalt für Fleischforschung Kulmbach* 1968

(8)   Authors' calculations and articles

(9)   Overbeck 'Das moderne Tierkörperbeseitigungswes'. *Die Fleisch-
        mehlindustrie* 1/1972 p. 6

(10)   Tierkörperbeseitigungsgesetz vom 1.2.1939 (RGBL I S. 187)
        (Carcass Disposal Act of 1 February 1939 (RGBL I page 187)

(11)   Durchführungsverordnungen zum Tierkörperbeseitigungsgesetz
        vom 23.2.1939 (RGBL L S. 322) vom 17.4.1939 (RGBL L S. 807)

(12) Gesetzentwurf Tierkörperbeseitigungsgesetz vom 11.2.1975
      (A Bill relating to the disposal of animal carcasses,
      11 February 1975)

(13) VDI Richtlinie (2590) *Anlagen zur Tierkörperverwertung*

(14) H.G. Korbitz and W. Brocke 'Geruchsquellen bei Tierkörperver-
      wertungsanstalten und techn, Massnahmen zur Vermeidung von
      Nachbarschaftsbelästigungen *Schriftenreihe der Landesanstalt
      für Immissions- und Bodennutzungsschutz des Landes*
      NRW Vol. k0/1968

(15) D. Strauch 'Flüssige Abfälle' *Die Fleischmehlindustrie*
      6/73 p 65 et seq

(16) H. Jung and G. Spennes 'Klärschlammkompostierung aus der Sicht
      der Abfallwirtschaft' Mensch - Wirtschaft - Technik
      Bauverlag Wiesbaden - Berlin 1974 p 305 et seq

(17) H. Schmidt 'Der Einsatz von Desinfektionsmitteln bei der
      Tierkörper- und Schlachtabfall beseitigung'
      Die Fleischwirtschaft 12/72 p 132

(18) Krupp Leaflets Hamburg - Hamburg

(19) E. Quellmalz *Die stoffliche Zusammensetzung von Geruchsstoffen*
      Vortrag Sept 1974 im VDI-Haus Düsseldorf

(20) *Schlacht- und Viehhofzeitung* 4/73 - p 149

(21) D. Strauch *Fragen der Hygiene bei organischen Düngern
      Garten organisch* 2/74

(22) K.H. Knoll *Hygienische Aspekte der Abfallbeseitigung*
      Bericht Abwassertechnische Vereinigung No. 26/1973

(23) Rheinstahl AG Essen Rheinstahl - Bühler Müllkompostierverfahren

OTHER LITERATURE

(24) Personal communications   Bundesminister für Ernahrung
Landwirtschaft und Forsten

(25)   Bericht des Arbeitskreises 'Tierische
Reststoffe' vom 15.11.1974

(26)   Abfallwirtschaftsprogramm 1975 der
Bundesregierung
- U II 6 - 530 021/I -

(27) Personal communications   Ministère de la Qualité de la vie et
de l'Environnement
Neuilly - France

(28) Personal communications   Handelsabteilung der Italienische
Botschaft in Bonn

(29) Personal communications   Ministère van Volksgezondheid en van
het Gezin
Brussels - Belgium

(30) Personal communications   Ministère de l'Agriculture
Luxembourg

(31) Personal communications   Ministerie van Landbouwen Visserij
Gravenhage - Netherlands

(32) Personal communications   Department of the Environment
London - England

(33) Personal communications   Ministry of the Environment
Agency of Environmental Protection
Copenhagen - Denmark

(34)   Giessener Berichte zum Umweltschutz
                                        Vol 1 - 5

(35)   *Umwelt-Zeitschrift*   VDI-Verlag Düsseldorf
                                        verschiedene Jahrgänge

(36)   K.H. Schiffers and   Mensch, Wirtschaft, Technik Bauverlag
       G. Spennes   G.m.b.H. Wiesbaden - Berlin

(37)   *Materialien zum Umweltprogramm*
                                        Bundesregierung 1971
                                        Verlag W. Kohlhammer

(38)   *Recycling in der Materialwirtsch*
                                        Vol 5 - Spiegel-Verlag